Structural Concrete Elements

Structural Concrete Elements

E. W. BENNETT

Reader in Civil Engineering,
Department of Civil Engineering,
University of Leeds

London
CHAPMAN AND HALL

First published 1973
by Chapman and Hall Ltd
11 New Fetter Lane, London EC4P 4EE
© 1973 E. W. Bennett
Typeset by Santype Ltd (Coldtype Division)
Salisbury, Wiltshire
Printed in Great Britain by
T. & A. Constable Ltd
Edinburgh

ISBN 0 412 09020 1

Distributed in the U.S.A.
by Halsted Press, a Division
of John Wiley & Sons, Inc., New York

Library of Congress Catalog Card Number 73-13380

To Dr Paul W. Abeles

in appreciation of twenty-five years
of criticism, advice and encouragement

Contents

Preface

In undergraduate and postgraduate teaching, the writer has increasingly felt the need for a single book treating the subject of structural concrete elements as a whole. Fundamental developments such as the use of prestressing, calculation of ultimate load and more recently limit state design, cannot now be adequately treated in chapters added to later editions of older classical textbooks on reinforced concrete. On the other hand the specialised and detailed works written for the practising engineer or research worker usually go far beyond the needs of the student.

It is, moreover, particularly unfortunate that the appearance of many separate books on reinforced and prestressed concrete should have created the erroneous impression of a basic distinction between the principles underlying the two forms of construction. In reality these principles are the same, so that there are conditions under which a prestressed member will behave in a manner closely resembling that of a reinforced concrete member, and vice versa. The work of the European Concrete Committee (C.E.B.) and International Prestressing Federation (F.I.P.) has done much to establish the essential unity of all types of structural concrete element, recognised in this country by the issue of a new unified Code of Practice for the Structural Use of Concrete. This unity cannot be ignored in the teaching of the next generation of engineers.

Several radical changes are at present taking place in the methods of calculation for structural concrete, and it is hoped that the value of this book will be increased by its being abreast of these developments. The Standard International (S.I.) metric system of units has been used, with the Imperial equivalents given in parenthesis. The notation is that agreed by the European Concrete Committee in May 1971 and since adopted as standard in both the British and American Codes of Practice. The principles of limit state design given in the former have been fully discussed and some of the significant recommendations of the new Code are given in the appropriate context.

The book covers the behaviour, theory and design of single structural concrete elements, but the rather complex interaction of such elements in statically indeterminate structures is not included. It should be a suitable textbook for an honours degree and provide the basic material for a more specialised postgraduate course. It is assumed that the reader will have some knowledge of the strength of materials and theory of structures, and although the relevant mechanical properties of concrete are described as they arise, the student will profit from the reading of a suitable textbook on the composition and properties of concrete.

The first part of the book gives an exposition of the mechanics of structural concrete elements, considering in each chapter a particular mode of behaviour which may be relevant in members either with or without prestress. The second part of the book provides an introduction to design calculations for students and beginners in practice. While it is hoped that this may afford some welcome guidance, the analytically minded student should bear in mind that mathematical calculations are only a tool in the art of design, and as such should have the right balance between simplicity and precision in approximating to the real behaviour of a structural concrete element.

The author is grateful to his wife and many colleagues who at various times have helped with the preparation and checking of the manuscript and diagrams. Particular thanks are due to Miss Susan Chapman and Mr Barry Firth for the able way in which they have done the major part of this unrewarding but important task.

Leeds   E.W.B.
January, 1973

Notation

(The complete form of some symbols is given in parenthesis)

A	Area, generally.
A_c	Area of concrete.
A_{ps}	Area of prestressing tendons (may be shortened to A_p).
A_s	Area of tension reinforcement *or*
	Area of non-prestressed tension reinforcement in a prestressed member
$A_s{}'$	Area of compression reinforcement.
A_{pe}	Equivalent area of prestressed steel having strength equal to the combined strength of the prestressing tendons and non-prestressed tension reinforcement.
A_{sl}	Area of longitudinal reinforcement provided for torsion.
A_{sv}	Cross-sectional area of the parallel legs of a link or links within a distance s (may be shortened to A_v).
a	Deflection.
a_{cr}	Distance from the point (crack) considered to the surface of the nearest longitudinal bar.

b	Breadth of section *or*
	Breadth of upper flange.
b_{inf}	Breadth of lower flange in I or T section.
b_t	Breadth of section at level of tension reinforcement.
b_w	Breadth of web of a member.
D_c	Density of concrete.
d	Effective depth of tension reinforcement.
d'	Depth to compression reinforcement.
d_b	Bar size.
E	Modulus of elasticity, generally.
E_c	Modulus of elasticity of concrete (static secant modulus)
E_{ce}	Effective (long-term) modulus of elasticity.
E_s	Modulus of elasticity of steel.
e	Eccentricity of a force with respect to the centroid of a section.
F	Load or force generally.
F_b	Anchorage load of reinforcement.
F_k	Characteristic load.
f	Direct stress or strength generally.
f_b,	Bond stress between steel and concrete.
f_c'	Cylinder strength of concrete (American Standard).
f_{ci}	Cube strength of concrete at initial transfer of prestress.
f_{cu}	(Characteristic) cube strength of concrete. (f_{cuk}).
f_{ct}	Direct tensile strength of concrete.
f_{cr}	Flexural tensile strength of concrete.
f_{sup}	Prestress in concrete at top of section ($f_{c\ sup\ p}$).
f_{inf}	Prestress in concrete at bottom of section ($f_{c\ inf\ p}$).
f_{cp}	Prestress in concrete at centroid of section ($f_{c\ m\ p}$).
$f_{c\ adm}$	Allowable compressive stress in concrete ($f_{c\ c\ adm}$).
$f_{t\ adm}$	Allowable tensile stress in concrete ($f_{c\ t\ adm}$).
$f_{c\ p\ adm}$	Allowable compressive stress in concrete at initial transfer of prestress. ($f_{c\ c\ p\ adm}$).
$f_{t\ p\ adm}$	Allowable tensile stress in concrete at initial transfer of prestress. ($f_{c\ t\ p\ adm}$).
f_{pb}	Tensile stress in tendons at (beam) failure.
f_{pe}	Effective prestress in tendons.
f_{pu}	(Characteristic) ultimate strength of tendons (f_{puk}).
f_s	Stress in reinforcement.
f_{sc}	Compressive stress in reinforcement.
f_{sb}	Tensile stress in reinforcement at beam failure.
f_{su}	(Characteristic) tensile strength of reinforcement (f_{suk}).
f_y	(Characteristic) yield stress of reinforcement. (f_{yk} *or* f_{syk}).
f_{yv}	(Characteristic) yield stress of link reinforcement. (f_{yvk}).
G	Shear modulus.
G_k	Characteristic dead load.

h	Overall depth of member.
h_f	Thickness of (upper) flange.
h_{inf}	Thickness of lower flange in I or T section.
h_{max}	Larger dimension of section.
h_{min}	Smaller dimension of section.
I	Second moment of area of a section, generally.
I_c	Second moment of area of uncracked concrete section.
I_e	Effective second moment of area for deflection calculations.
I_r	Second moment of area of cracked concrete section.
I_s	Second moment of area of steel section.
I_x	Second moment of area about the major (horizontal) axis of bending.
I_y	Second moment of area about the minor (vertical) axis of bending.
i	Radius of gyration.
$\left.\begin{array}{l}K\\k\end{array}\right\}$	Constants, having dimensions.
l	Effective span.
M	Bending moment, generally.
M_a	Increased moment in column.
M_{cr}	Cracking moment.
M_d	Design moment (serviceability limit state).
M_g	Bending moment due to dead load.
M_o	Moment necessary to produce zero stress in the concrete of a prestressed beam.
M_r	Moment causing cracking of concrete.
M_u	Ultimate moment.
M_{ud}	Design moment for ultimate limit state.
M_x, M_y	Moments about the major and minor axes of a column.
N	Force normal to a section, generally.
N_{bal}	Load on a column for balanced condition.
P	Prestressing force.
Q_k	Characteristic imposed load.
r	Internal radius of bend.
S	First moment, about the axis of bending, of the area outside a line, parallel to the axis, through a given point.
s	Standard deviation *or* Spacing of links (s_v)
T	Torsional moment generally.
T_u	Ultimate torsional moment.
T_{uo}	Ultimate torsional moment in the absence of shear.
t	Time.
u	Perimeter.

V	Shear force, generally.
V_c	Ultimate shear resistance of concrete.
V_{co}	Ultimate shear resistance of concrete in a section uncracked in flexure.
V_{cr}	Ultimate shear resistance of concrete in a section cracked in flexure.
V_u	Ultimate shear force.
V_{ud}	Design shear force for ultimate limit state.
V_{uo}	Ultimate shear in the absence of torsion.
v	Shear stress, generally.
v_c	Ultimate shear stress in concrete.
v_t	Torsional shear stress.
v_{tu}	Ultimate torsional shear stress.
v_u	Ultimate shear stress.
W_k	Characteristic wind load.
x	Co-ordinate; *or* Neutral axis depth
x_1	Smaller dimension of a link.
y	Co-ordinate.
y_o	Half depth of an anchorage block.
y_{po}	Half loaded area of anchorage block.
y_1	Larger dimension of a link.
z	Co-ordinate *or* Lever arm.
α	Angle, ratio or dimensionless coefficient, generally *or* Modular ration (α_e) *or* Angle of inclination of shear reinforcement (α_v)
β	Angle, ratio or dimensionless coefficient, generally.
γ_m	Partial factor for materials.
γ_f	Partial factor for forces.
δ	Coefficient of variation.
ϵ	Strain, generally.
ϵ_c	Strain in concrete.
ϵ_{cs}	Limiting shrinkage strain in concrete.
ϵ_{cu}	Ultimate compressive strain in concrete.
ϵ_s	Strain in steel.
ζ	Dimensionless coefficient.
η	Reduction factor for loss of prestress.
θ	Angle of inclination of tendons.
ν	Poisson's ratio.
ξ	Neutral axis depth ratio $\left(\dfrac{x}{d} \text{ or } \dfrac{x}{h}\right)$
ρ	Geometrical ratio of reinforcement.
ϕ	Creep coefficient *or* Capacity reduction factor.
ψ	Dimensionless coefficient *or* probability.

Part I
Mechanics of Structural Concrete Elements

1

Uncracked elements with linear deformation

Principles and assumptions

Concrete structures may be of plain, reinforced or prestressed concrete. The use of plain concrete is generally confined to mass concrete and paving, whereas individual structural elements are either reinforced by steel or prestressed order to improve their mechanical properties.

The tensile strength of concrete is low in relation to the compressive strength, the ratio being rarely more than one-tenth, and decreasing as the strength increases. Tensile cracking is therefore only absent from plain or reinforced concrete members which are very lightly loaded, or which are designed to function mainly in compression. Reinforced concrete liquid-retaining structures come into the former class, while the latter includes gravity dams, retaining walls, columns and arch ribs. A further important class comprises prestressed structures in which controlled internal forces are used to maintain a state of compression in those parts of the concrete which would otherwise be in tension under the effect of the external load. In considering the behaviour of uncracked concrete elements one is therefore dealing with the full range of working load of most prestressed members, but with relatively small loads on reinforced concrete members in tension or flexure.

The initial deformation of uncracked concrete elements is usually assumed to be linearly proportional to the load. The complete stress-strain characteristic of concrete in compression is not linear, but it will be seen from an examination of Fig. 1.1 that the assumption is valid provided that the stress does not exceed about one third of the compressive strength. The tensile stress-strain characteristic appears to be almost linear up to the point of failure.

When concrete is maintained in a state of constant stress it continues to undergo a further deformation or 'creep' which approaches a limiting value after a period of time. Since the creep of concrete is approximately proportional to the stress up to about one third of the compressive strength the deformation of concrete structures under sustained loading may also be treated as linear as illustrated in Fig. 1.1 with the adoption of a lower stress-strain modulus (E_c). This is the justification for the value of 15 commonly used in conventional reinforced concrete calculations for the modular ratio of steel to concrete

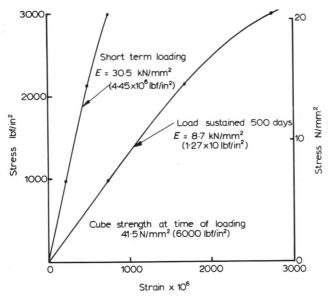

Figure 1.1   Stress-strain curves for concrete
(Proportions 1:1:2, water-cement ratio 0.375)

($\alpha = E_s/E_c$). If E_s is 200 kN/mm^2 (29 x 10^6 lbf/in^2) the corresponding value of E_c is about 13 kN/mm^2 (2 x 10^6 lbf/in^2), that is between one third and one half of the value for a short term load. Not all the deflexion due to sustained load is recovered on removal of the load, however, so that the behaviour although approximately linear cannot be termed elastic.

It is usual, and both theoretically and experimentally justifiable to assume the distribution of strain across the section of a member composed of isotropic material to be linear, provided that the depth of the section is small in relation to the length of the member. When the member is of concrete with longitudinal steel reinforcement the linear distribution of strain may still be assumed provided that there is no relative movement between the reinforcing bars and the surrounding concrete, which further assumption can be justified experimentally in the absence of tensile cracking. The stress in the reinforcement will, however, differ from that in the adjacent concrete on account of the different moduli of the two materials, as shown in the following example.

Reinforced concrete member under axial load

The calculation of stresses in reinforced concrete members is based on two conditions, the first of which is that the internal and external forces and moments must be in equilibrium, and the second that the strain is linearly distributed and related to the stress in the material at each point by the appropriate modulus of elasticity. In the simple example of an axially loaded reinforced concrete member each of these conditions is represented by one equation.

Consider a member subject to an axial compressive force N in which the cross-sectional area of concrete is A_c and that of the reinforcement is A_s (Fig. 1.2). If the compressive stresses are f_s and f_c in the steel and concrete respectively, since the axial forces are in equilibrium

$$N = f_c A_c + f_s A_s.$$
                                                                    1.1

Since the longitudinal strain is assumed to be equal at all points on a cross section

$$\frac{f_s}{E_s} = \frac{f_c}{E_c}$$

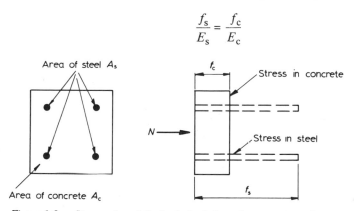

Figure 1.2   Stresses in axially loaded reinforced concrete member

Therefore

$$f_s = \frac{E_s}{E_c} \cdot f_c = \alpha f_c. \qquad\qquad 1.2$$

From (1.1) and (1.2)

$$N = f_c(A_c + \alpha A_s).$$

It will be noted that the term $A_c + \alpha A_s$ represents the cross sectional area of a plain concrete member which would develop the same stress f_c under an axial load N.

This equivalent concrete section, sometimes termed the 'transformed section', is a useful concept in the analysis of structural concrete, as indeed of all composite members.

In a reinforced concrete member of rectangular cross section with dimensions b and h as in Fig. 1.3 its area may be written in the form

$$A_c + \alpha A_s = (bh - A_s) + \alpha A_s$$

$$= bh + (\alpha - 1) A_s.$$

that is as the nominal area of cross-section to which is added $(\alpha - 1)$ times the area of the reinforcement.

It should be explained that this form of calculation is no longer normally used for compression, having been superseded by the method given in Chapter 4.

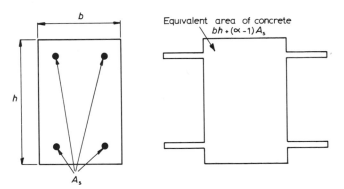

Figure 1.3   Equivalent ('transformed') concrete section

One instance in which it is employed for tension, however, is the analysis of stresses relating to the resistance to cracking of parts of water-retaining structures in direct tension according to the British Standard Code of Practice CP 2007 (1960). An example of this follows.

EXAMPLE

The maximum circumferential force in a cylindrical reinforced concrete water tank is 306 kN per metre (21 000 lbf per ft) of height and the thickness of the wall of the tank is 190 mm (7½ in). Calculate the reinforcement required if a concrete of nominal mix 1 : 1.6 : 3.2 is to be used.

The Code of Practice CP2007 specifies in Table 2 that the direct tensile stress for the above nominal mix must not exceed 1.31 N/mm^2 (190 lbf/in^2) to ensure adequate resistance to cracking.

$$\text{Equivalent area of concrete required} = \frac{306 \times 10^3}{1.31} = 234\ 000 \text{ mm}^2 \text{ per m height}$$

$$(110.5 \text{ in}^2 \text{ per ft height})$$

$$\text{Nominal area of concrete} = 190 \times 1000 = 190\ 000 \text{ mm per m height}$$

$$(90 \text{ in}^2 \text{ per ft height})$$

Equivalent additional area of concrete to be provided by steel

$$= 234\ 000 - 190\ 000 = 44\ 000 \text{ mm}^2 \text{ per m height}$$

$$(20.5 \text{ in}^2 \text{ per ft height})$$

$$\text{Area of steel required} = \frac{44\ 000}{15 - 1} = 3140 \text{ mm}^2 \text{ per m height}$$

$$(1.465 \text{ in}^2 \text{ per ft height})$$

12 mm bars at 70 mm centres provide 3230 mm^2 per m
(½ in bars at 3 in centres provide 1.570 in^2 per ft)

The tensile stress in the steel in uncracked reinforced concrete members is relatively low and in this example the stress corresponding to 1.31 N/mm^2 (190 lbf/in^2) in the concrete is only 1.31 x 15 = 19.6 N/mm^2 (2850 lbf/in^2). Since the yield stress of steel is at least 275 N/mm^2 (40 000 lbf/in^2) it is clearly not possible to use reinforcement economically in an uncracked member in tension.

Reinforced concrete member subject to pure bending

The flexural stresses in an uncracked reinforced concrete member may conveniently be calculated by the conventional bending formulae, using the equivalent concrete section.

The equivalent area of the section shown in Fig. 1.4 is

$$A = bh + (\alpha - 1)A_s.$$

The position of the centroidal axis, which is the neutral axis of bending is

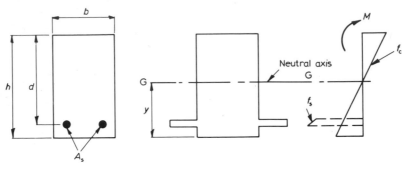

Section of member　　　　Equivalent concrete section　　　　Stresses

Figure 1.4　　Bending stresses in uncracked reinforced concrete member

defined by the dimension $\bar{y}$, calculated from the moment of area about the bottom of the section

$$\bar{y} = \frac{bh \cdot h/2 + (\alpha - 1)\,A_s\,(h - d)}{bh + (\alpha - 1)A_s}.$$

The second moment of area, I, ('Moment of inertia') of the section about its centroidal axis is then given by

$$I = \frac{bh^3}{12} + bh\left(\frac{h}{2} - \bar{y}\right)^2 + (\alpha - 1)A_s\,(\bar{y} - h + d)^2.$$

The stress in the concrete, f_c, then has the value $M\bar{y}/I$ at the bottom and $M(h - \bar{y})/I$ at the top of a section at which the moment is M. The stress in the steel reinforcement is

$$f_s = \frac{\alpha M}{I}\,(\bar{y} - h + d).$$

If preferred, the stresses may be calculated by the use of the section moduli, i.e.

$$f_c = \frac{M}{Z_c} \text{ and } f_s = \frac{M}{Z_s},$$

where　$Z_c = \dfrac{I}{\bar{y}}$　and　$\dfrac{I}{h - \bar{y}}$　for the bottom and top respectively,

and

$$Z_s = \frac{I}{\alpha(\bar{y} - h + d)}.$$

The initial uncracked condition of a reinforced concrete member in flexure terminates at a low load as soon as the tensile strength of the concrete is exceeded. The above calculation may therefore be used to estimate the load at which tensile cracks may be expected to appear, on the assumption that this is determined by a critical value of the stress on the tensile face of the member. In point of fact the cracking load is considerably affected by the extent of shrinkage of the concrete, as discussed in Chapter 3. An example of a calculation neglecting shrinkage is given below.

EXAMPLE

The rectangular cross section of a reinforced concrete beam is 250 mm (10 in) wide and 450 mm (18 in) deep and the tensile reinforcement consists of three 20 mm (¾ in) bars located 400 mm (16 in) from the top. If the concrete cracks when the tensile stress at the soffit is 4.15 N/mm² (600 lbf/in²) calculate the cracking moment and the corresponding stress in the reinforcement.

Area of equivalent concrete section

$$A = 250 \times 450 + (15 - 1) \times 942 = 125\ 200\ \text{mm}^2\ (198.5\ \text{in}^2).$$

Distance of centroid from bottom

$$\bar{y} = \frac{112\ 500 \times 225 + 13\ 200 \times 50}{125\ 700} = 207\ \text{mm}\ (8.3\ \text{in}).$$

Second moment of area about centroid

$$I = \frac{112\ 500 \times 450^2}{12} + 112\ 500\ (225{-}207)^2 + 13\ 200\ (207{-}50)^2$$

$$= 2270 \times 10^6\ \text{mm}^4\ (5678\ \text{in}^4).$$

Section modulus for concrete at bottom

$$Z_c = \frac{2270 \times 10^6}{207} = 11.0 \times 10^6\ \text{mm}^3\ (682\ \text{in}^4).$$

Section modulus for steel

$$Z_s = \frac{2270 \times 10^6}{15 \times 157} = 0.965 \times 10^6\ \text{mm}^3\ (60\ \text{in}^3).$$

Cracking moment

$$M = 4.15 \times 11.0 \times 10^6 = 45.7\ \text{kN m}\ (410\ 000\ \text{lbf. in}).$$

Stress in steel immediately before cracking

$$f_s = \frac{45.7 \times 10^6}{0.965 \times 10^6} = 47.3\ \text{N/mm}^2\ (6850\ \text{lbf/in}^2).$$

As in the previous example, the stress in the reinforcement is uneconomically low.

Combined axial force and bending

The calculated stresses due to an axial force may be combined with those resulting from a bending moment when the two actions occur together, as for example, when there is an eccentric load on a member, which is equivalent to an axial force and a moment. In the absence of cracking the use of the equivalent concrete section is again appropriate.

In the diagram given as Fig. 1.5 a reinforced concrete member is acted on by a compressive force N at eccentricity e. The compressive stress in the concrete at a distance e_c from the centroidal axis is given by

$$f_c = \frac{N}{A} + \frac{N}{I} \cdot y_c.$$

The sign of y_c is positive for points on the same side of the centroidal axis as the line of action of the force N. A tensile stress on the opposite side of the axis will be denoted by a negative value of f_c.

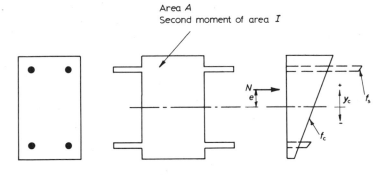

Area A
Second moment of area I

Section of member   Equivalent concrete section   Stresses

Figure 1.5   Combined axial and bending stresses in uncracked reinforced concrete member

The capacity of a given section to resist various combinations of load and moment, without exceeding specified permissible stresses in compression and tension, may be graphically represented by a load-moment diagram, as in Fig. 1.6. If the compressive stress in the concrete in bending must not exceed a permissible value $f_{c\,adm}$, the above relationship for the compressive face $(e_c = y_{sup})$ yields the condition

$$\left(\frac{1}{A}\right) N + \left(\frac{y_{sup}}{I}\right) N \leqslant f_{c\,adm}.$$

The limiting case in which the permissible stress is just attained is represented by the straight line shown on the diagram of N against Ne in Fig. 1.6.
Similarly if the flexural tensile stress is limited to a permissible value $f_{ct\,adm}$ a

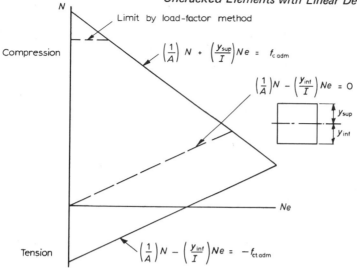

Figure 1.6   Load-moment diagram for uncracked reinforced concrete member

second stress condition is obtained for the tensile face $(e_c = -y_{inf})$

$$-\left(\frac{1}{A}\right) N + \left(\frac{y_{inf}}{I}\right) N_e \lessgtr f_{ct\ adm}$$

Here the sign on the left hand side of the expression has been reversed so that the permissible tensile stress on the right hand side may be regarded as positive.

The limit of this second condition is also shown as a straight line on the load-moment diagram, while the broken line shows the special case in which no tensile stress is permitted $(f_{ct\ adm} = 0)$.

The given section is therefore seen to be capable of resisting all combinations of load and moment denoted by points lying within the triangle bounded by the vertical axis and the two lines representing the limiting stress conditions.

The load-moment diagram is widely employed for the design of reinforced concrete sections, and the linear method here described was in fact permitted for the design of columns subject to direct load and bending according to the Code of Practice CP 114: 1957 (para. 322 (c)) with the proviso that the permissible axial load should not exceed the value calculated by the load factor method as described in Chapter 4.

Composite members of steel and concrete

Reinforced concrete may be regarded as a special case of composite construction in steel and concrete in which the cross-sectional dimensions of the steel elements are small in relation to the concrete section. It is therefore permissible and customary to neglect any gradient of stress due to bending of the reinforcing bars and to consider the stress in them to be uniform, even, for example when the tensile reinforcement in a beam is placed in more than one layer. This assumption is clearly inappropriate, however, when a composite section consists largely of steel, but it need not be made if the calculation of bending stresses is based on the equivalent section, either of steel or of concrete. An example will serve to illustrate this.

EXAMPLE

Calculate the maximum stress in the steel and concrete when the composite steel and concrete beam shown in Figure 1.7 is subject to a bending moment of 330 kN m (1300 tonf in).

The British Standard Code for Composite Construction, CP 117: Part 1: 1965 recommends that the long-term modular ratio of 15 should be used for the calculation of stresses. On this basis the dimensional properties of the equivalent steel section are found to be

$$A = 17\ 700 \text{ mm}^2 \ (27.44 \text{ in}^2),$$

$$I = 835 \times 10^6 \text{ mm}^4 \ (2004.2 \text{ in}^4).$$

Distance of centroid from bottom $= 356$ mm (14.03 in).

Stress in steel at soffit

$$f_s = 330 \times 10^3 \times 10^3 \text{ N mm} \times \frac{356 \text{ mm}}{835 \times 10^6 \text{ mm}^4} = 140 \text{ N/mm}^2 \ (9.1 \text{ tonf/in}^2).$$

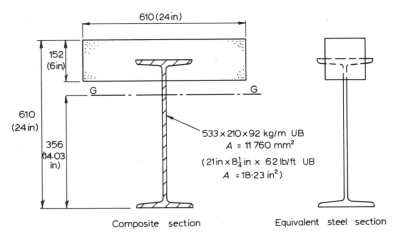

Figure 1.7   Composite beam of structural steel and concrete

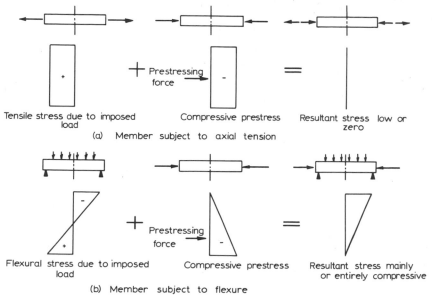

Figure 1.8   Use of compressive prestress to counteract tensile stress

Stress in concrete at top of slab

$$f_c = \frac{330 \times 10^6 \ (610 - 356)}{835 \times 10^6 \times 15} = 6.70 \ \text{N/mm}^2 \ (965 \ \text{lbf/in}).$$

Compressive prestressing

The relative weakness of concrete in tension has already been noted, as also has the consequently limited tensile or flexural load that can be withstood by reinforced concrete members without cracking, and the low level of stress in the reinforcement. Conventional tensile reinforcement can in fact only be used economically when a certain degree of cracking of the concrete is permitted. Prestressing presents an alternative solution to the problem posed by the tensile weakness of concrete and enables members to be maintained in an uncracked condition over a very much greater range of applied load, with the attendant advantages of reduced deflexion and increased durability.

The principle of prestressing as applied to members subject to axial tension and pure flexure respectively, is simply illustrated in Fig. 1.8. The tensile stress resulting from the axial load on the member is counteracted by a compressive prestress produced by a prestressing force also acting along the axis of the member. Since the deformation is linear the principle of superposition is valid and the resultant stress is the algebraic sum of the stress due to the load and the prestress. In Fig. 1.8(b) the flexural stress due to the imposed load on the beam is partly tensile and partly compressive; the former is cancelled by the prestress produced by an eccentric force along the beam, while the increased resultant compressive stress in the upper part of the beam is easily accommodated by the concrete.

Prestressing techniques

The majority of prestressed concrete members require a prestressing force acting parallel or inclined at a small angle to their axis. Although the direct use of external jacks is feasible under certain conditions the most common prestressing technique involves the use of tensioned flexible high strength steel elements, usually known as tendons, which pass along the length of the member and transmit a compressive reaction to it at the ends. The tendons may consist of individual hard drawn wires, cables of hard drawn wires, either separate or stranded, or bars of high-strength alloy steel, and the techniques fall into two main classes, termed pre-tensioning and post-tensioning. Pre-tensioned tendons are bonded to the concrete like ordinary reinforcement, and the prestressing force is transmitted by the bond over a certain distance (the 'transmission length') at each end. In order to cast a beam with pretensioned steel it is necessary to maintain the tendons in tension temporarily by means of external anchorages until the concrete has hardened, at which stage the prestress is transferred to the concrete by slowly releasing the tension from these anchorages.

Post-tensioned tendons are tensioned after the concrete in the beam has hardened and are located in ducts through the member or sometimes externally alongside it. The process requires the use of a portable jack for tensioning, and permanent anchorages to transmit the tensile force at each end. It has the advantage of enabling the line of the prestressing force to be curved, which is difficult in pre-tensioning, and permits prestressing on the site without the need for heavy temporary anchorages. However, the anchorages required for each tendon are a source of expense, so that post-tensioning is generally employed for larger members and pre-tensioning for the factory production of members of small to medium size in rows on long beds.

Limits of stress

The magnitude of prestress that can be used in a particular member will be governed by the fact that the concrete must not be overstressed in compression or in tension. The limiting factor is therefore the resultant stress in the concrete under the most unfavourable combinations of prestress and loading; the actual values of the prestress and loading stress which contribute to the resultant are not in themselves significant.

The extreme values of resultant stress in the concrete occur under the maximum and minimum imposed loads. The latter may be zero, it may correspond to a permanent dead load, or in some instances, it may be a negative value brought about by temporary conditions during erection. The critical stress conditions for a simply supported beam are shown in Fig. 1.9 from which it will be seen that the maximum compressive stress occurs at the bottom of the section under the minimum load and at the top under the maximum load while the reverse is true of the maximum tensile stress, or minimum compressive stress if no tensile stress occurs.

Since it is the resultant stress rather than the prestress which determines whether the concrete is overstressed, it is possible for a section to accommodate a relatively large minimum load by a correspondingly large prestress. This is an important advantage of prestressed members and its limit is imposed mainly by practical considerations.

Two further factors have to be considered in practice. The first is that the prestressing force does not remain constant but diminishes with time to approach a final value which may be as low as 70% of the initial force. This loss

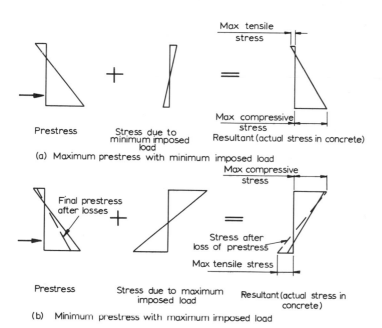

Figure 1.9   Critical stresses in a prestressed beam

of prestress, which will be more fully discussed in Chapter 3, has the important implication that the resultant stresses under the maximum imposed load are greater when it has occurred, as indicated by the broken lines in Fig. 1.9(b).

The use of high strength steel for prestressing tendons is dictated by the need to minimise the proportional reduction of the prestress, and for the same reason it is normal practice to use a high initial prestress in the steel, of the order of 70% of the tensile strength. The final prestress will be about one half of the strength and although the effect of the imposed load will be to increase the stress in the steel it can readily be shown by calculation that the total stress will remain well below the initial value.

The second practical consideration arises from the fact that the permissible stresses in the concrete are related to the strength of the concrete, which may increase appreciably between the time at which the prestress is applied and the time at which the maximum load first occurs. An economical design will therefore require different permissible stresses for the two critical stress conditions represented in Fig. 1.9(a) and (b), and the various factors governing the specification of these stresses will be discussed in Chapter 7.

Calculation of prestress and bending stress

The method of calculation of the prestress is the same as that already shown for an uncracked section subject to an eccentric load (p. 8).

Using the notation indicated in Fig. 1.10 the compressive stress at the bottom of the section is given by

$$f_{\inf} = \frac{P}{A} + \frac{Pey_{\inf}}{I}$$

$$= \frac{P}{A}\left(1 + \frac{y_{\inf}e}{i^2}\right)$$

where $I = Ai^2$.

Similarly at the top of the section there will be a tensile stress given by

$$f_{\sup} = \frac{P}{A}\left(\frac{y_{\sup}e}{i^2} - 1\right).$$

The bending stress due to a moment M will be $\dfrac{M}{Z_{\inf}}$ at the bottom and $\dfrac{M}{Z_{\sup}}$ at the top where $Z_{\inf} = \dfrac{I}{y_{\inf}}$ and $Z_{\sup} = \dfrac{I}{y_{\sup}}$.

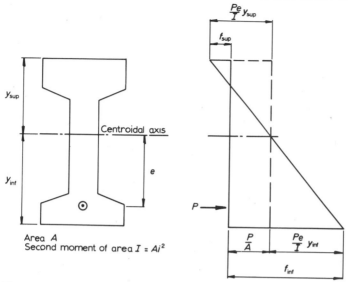

Figure 1.10   Calculation of prestress

The prestressing force is the product of the prestress in the steel and the cross sectional area of tensioned steel. The prestress is the same as the initial steel stress applied when the tendons are all post-tensioned in a single operation, but when they are pretensioned the initial stress is reduced by the elastic deformation of the concrete to a lower value, known as the prestress at the stage of transfer. When several tendons in a member are prestressed consecutively there is a reduction of stress in the first tendons to be stressed and anchored owing to elastic deformation of the concrete as the later tendons are stressed.

Since the area of steel, both in tendons and in ordinary reinforcement is usually small in prestressed concrete members, it is often satisfactory to calculate the nominal dimensional properties of the section ($Ay_{\inf}$ and I) ignoring steel, cable ducts, etc. When, however, the maximum accuracy is required it is necessary to take account of the fact that the section of a member resisting the prestressing force is slightly different from the section which resists externally applied loads.

Considering the prestress alone, there is a state of equilibrium in which the tensioned steel is the active element providing the prestressing force, while the concrete section, together with any supplementary untensioned reinforcement acts passively. In calculating the dimensional properties of the section, the area occupied by pre-tensioned steel should therefore be deducted, while if the steel is post-tensioned the cross sectional area of the ducts will not of course be

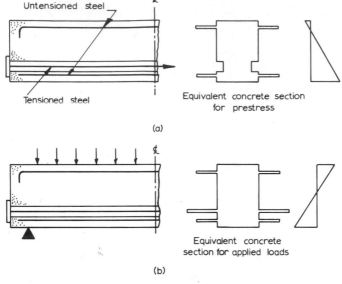

Figure 1.11   Sections for calculation of stresses

included. The section used for computation of the prestress is illustrated
diagramatically in Fig. 1.11(a).

When, however, the member is deformed by externally imposed loads, the
tensioned steel will also contribute to the passive resistance of the section, which
should therefore be calculated to include the equivalent area of this steel, that is
its area multiplied by the modular ratio as indicated in Fig. 1.11(b).

The calculation of the dimensional properties of prestressed concrete sections
by the more accurate method described above is somewhat laborious and a
systematic approach is essential. An example of a tabulated desk machine
calculation for a section with pre-tensioned tendons is given as Appendix A, and
the operation is readily programmed for a digital computer.

Calculation of prestress at transfer from initial stress in steel

The above method has two disadvantages. The first is that the prestress in the
steel tendons at transfer must be calculated or estimated and its value is
dependent on the elastic deformation and hence on the prestress in the concrete.
The second disadvantage lies in the necessity of calculating two sets of
dimensional properties of the section.

A direct calculation of the prestress at transfer in a member with
pre-tensioned steel may be made for the initial stress in the steel, which is
usually known fairly accurately, using only the equivalent section for applied

loads. This permits a result of maximum accuracy with a reduced amount of computation, as the same section is used to calculate both the prestress and the stresses due to imposed loads.

When a member with pre-tensioned steel has been cast it is in equilibrium under the external action of the prestressing force P at each end with an initial stress f_{pi} in the steel and zero stress in the concrete as in Fig. 1.12(a). After transfer of the prestress to the concrete, the stresses in both the steel and the concrete may be calculated by superposing on the existing stresses the stresses set up by removing the external tensile forces P, which is equivalent to applying compressive forces of equal value to the complete section including the equivalent area of tensioned steel. This is shown in Fig. 1.12(b).

EXAMPLE

The cross section of a prestressed concrete beam 7.60 m (25 ft) long is shown in Fig. 1.13. If the wires are tensioned to an initial stress of 1100 N/mm² (160 000 lbf/in²) calculate the resultant stress at transfer at the midspan point, (a) in the concrete in the soffit and (b) in the bottom wires, assuming that the modulus of elasticity of the concrete is 27 500 N/mm² (4.0 × 10⁶ lbf/in²) and the modulus of elasticity of the steel 193 000 N/mm² (28.0 × 10⁶ lbf/in²).

The modular ratio is $\dfrac{193\ 000}{27\ 500} = 7$ and the dimensional properties of the equivalent section are

A = 38 200 mm² (59.26 in²)

y_{inf} = 145 mm (5.70 in),

I = 443 × 10⁶ mm⁴ (1065.0 in⁴),

i^2 = 11 600 mm² (17.98 in²),

e = 77 mm (3.03 in).

Initial prestressing force = 1100 × 12 × 19.6 = 259 000 N (60 200 lbf).

Compressive prestress at soffit at transfer

$$f_{inf} = \frac{259\ 000}{38\ 200}\left(1 + \frac{145 \times 77}{11\ 600}\right) = 13.30 \text{ N/mm}^2 \quad (1990 \text{ lbf/in}^2).$$

Compressive prestress at level of bottom wires

$$= \frac{259\ 000}{38\ 200}\left(1 + \frac{120 \times 77}{11\ 600}\right) = 12.20 \text{ N/mm}^2 \quad (1820 \text{ lbf/in}^2).$$

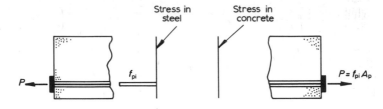

(a)   Equilibrium under external forces

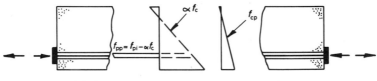

(b)   Balancing forces opposed by stresses in
equivalent section

Figure 1.12   Calculation of prestress from initial stress in pretensioned steel

Taking the density of concrete as 2400 kg/m³ (150 lb/ft³) the self-weight of the beam

$$= 38\ 200 \times 10^{-6} \times 2400 \times 9.81 = 900 \text{ N/m} \ (61.5 \text{ lb/ft}).$$

Moment at midspan due to self-weight

$$= 900 \times 7.60^2 \times 0.125 = 6500 \text{ Nm} \ (57\ 600 \text{ lbf/in}).$$

Resultant compressive stress at soffit

$$= 13.30 - 6500 \times 10^3 \times \frac{145}{443 \times 10^6} = 11.18 \text{ N/mm}^2 \ (1680 \text{ lbf/in}^2).$$

Compressive stress in concrete at level of bottom wires

$$= 12.20 - 6500 \times 10^3 \times \frac{120}{443 \times 10^6} = 10.44 \text{ N/mm}^2 \ (1570 \text{ lbf/in}^2).$$

Tensile stress in bottom wires

$$= 1100 - 10.44 \times 7 = 1027 \text{ N/mm}^2 \ (149\ 000 \text{ lbf/in}^2).$$

It is seen from this example that the increase of stress in the steel due to bending, like that in uncracked reinforced concrete members, is small.

The above method can be applied to members with post-tensioned steel in order to retain the advantage of using only the complete equivalent section in calculations. If the prestress is applied to the steel in one operation the initial prestressing force P is also the prestressing force at transfer. If βP is the corresponding imaginary force that would cause the same prestress by acting on the complete equivalent section, β can be obtained as follows

Compressive stress in concrete at level of steel

$$= \frac{\beta P}{A} \left(1 + \frac{e}{i^2}\right).$$

Hence force in steel at transfer

$$= \beta P - \frac{\alpha A_p \beta P}{A} \left(1 + \frac{e^2}{i^2}\right) = P,$$

in which α is the modular ratio and A_p the area of prestressed steel.

Hence

$$\beta = \frac{A i^2}{A i^2 - \alpha A_p(i^2 + e^2)}.$$

When the steel is tensioned by a series of consecutive jacking operations, for example in large members with multiple tendons or in the various single-wire systems there will be a progressive shortening due to compression and a consequent uneven loss of stress in the steel. It has been shown that the average loss of stress does not exceed about one half of that for an equivalent beam with pre-tensioned steel so that β may be expected to lie between the value given by the above expression and the mean of this value and unity, depending on the number of stages in which the prestress is applied.

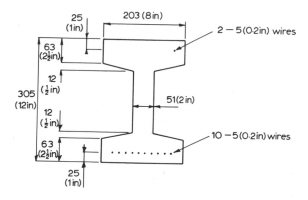

Figure 1.13   Example of calculation of prestress

Composite members of prestressed and added concrete

It is often possible to economise by the use of composite members in which a prestressed concrete element, smaller than would be necessary for the full load, is increased in stiffness by the addition of more concrete at relatively low cost. In the most usual form of composite member the prestressed concrete element is pre-cast and the member completed by adding the remainder of the concrete *in situ*. Common examples are bridge decks, either of T section which the web and bottom flange is prestressed and the top flange added to form the deck, or in which a slab is formed by pre-cast beams of inverted T section acting as permanent shuttering for the added concrete. In each example, the added concrete, which usually includes some reinforcement, provides lateral stiffness.

The calculation of the stresses is based on the principle that a single integral member is formed when the added concrete has hardened so that the additional stresses due to any subsequent loadings are obtained using the increased dimensional properties of the entire composite section. In the prestressed element these stresses are superposed on the stresses existing before the added concrete hardens. This assumes that no slip occurs at the interface between the prestressed and the added concrete, and care is necessary that the condition is fulfilled.

The added concrete, being younger and sometimes of inferior quality, may have a lower modulus of elasticity than that in the pre-cast part of the member. When this is so the dimensional properties should be calculated for an equivalent section in terms of the concrete in either part of the member.

EXAMPLE

The cross-section of a pre-cast prestressed beam of span 8.5 mm (28 ft) is shown in Fig. 1.14. The modulus of elasticity of the concrete is 35 kN/mm² (5 × 10⁶ lbf/in²) and the prestressing force is 20 kN (4 500 lbf) in each wire at transfer. Calculate the prestress in the concrete at transfer. If, after a 20 per cent loss of prestress, additional concrete with a modulus of elasticity of 28 kN/mm² (4 × 10⁶ lbf/in²) is placed to form a composite slab 460 mm (18 in) deep while the beam is supported on its bearings, find the maximum uniformly distributed load that can be imposed if no tensile stress is to occur at the soffit of the beam.

Prestressing force at transfer,

$$P = 23 \times 20 = 460 \text{ kN (103 500 lbf)}.$$

Eccentricity of prestressing force

$$e = \frac{20 \times 115 - 3 \times 215}{23} = 72 \text{ mm (2.80 in)}.$$

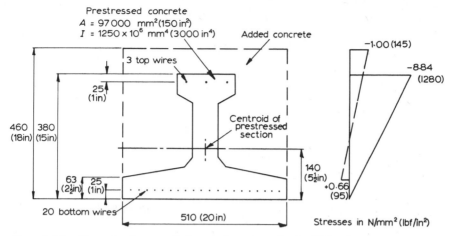

Figure 1.14 Example of composite beam of prestressed and additional concrete

Compressive prestress at bottom

$$f_{inf} = \frac{460 \times 10^3}{97\,000}\left(1 + \frac{140 \times 72}{12\,900}\right) = 8.46 \text{ N/mm}^2 \ (1220 \text{ lbf/in}^2).$$

Tensile prestress at top,

$$f_{sup} = \frac{460 \times 10^3}{97\,000}\left(\frac{240 \times 72}{12\,900} - 1\right) = 1.59 \text{ N/mm}^2 \ (230 \text{ lbf/in}^2).$$

Modular ratio of added concrete to prestressed concrete $= \dfrac{28}{35} = 0.80$

The equivalent composite section (in terms of prestressed concrete) is obtained by multiplying the area of the prestressed concrete by 1.0 and the area of the added concrete by 0.8. In the present example this is more conveniently done by multiplying the gross area by 0.8 and adding 0.2 times the area of the prestressed section.

$$A = 510 \times 460 \times 0.8 + 97\,000 \times 0.2 = 207\,000 \text{ mm}^2 \ (318 \text{ in}^2).$$

$$y_{inf} = 230 - \frac{19\,400\,(230 - 140)}{207\,000} = 222 \text{ mm} \ (8.76 \text{ in}).$$

$$I = 187\,600 \times \frac{460^2}{12} + 1746\,000(230 - 140) + 1250 \times 10^6 \times 0.2 - 207\,000 \times 8^2$$

$$= 3694 \times 10^6 \text{ mm}^4 \ (8713 \text{ in}^4).$$

$$Z_{inf} = \frac{3694 \times 10^6}{222} = 16.6 \times 10^6 \text{ mm}^3 \text{ (1005 in}^3\text{)}.$$

Weight of beam and added concrete

$$= 2400 \times 510 \times 460 \times 10^{-6} \times 9.81 = 5520 \text{ N/m (378 lbf/ft)}.$$

Moment due to weight of beam and added concrete

$$= 5520 \times 8.5^2 \times 0.125 = 49\,900 \text{ N m (440 000 lbf in)}.$$

Resultant compressive stress at soffit (calculated for prestressed beam section supporting self weight and weight of added concrete)

$$= 0.8 \times 8.46 - 49\,900 \times \frac{140}{1250 \times 10^6} = 1.17 \text{ N/mm}^2 \text{ (170 lbf/in}^2\text{)}.$$

Permissible imposed moment

$$= 1.17 \times 16.6 \times 10^6 = 19.4 \text{ kN m (178 000 lbf/in}^2\text{)}.$$

Permissible uniformly distributed imposed load

$$= \frac{19.4}{8.5^2 \times 0.125} = 2.15 \text{ kN/m (151 lbf/ft)}.$$

The distribution of stress may easily be calculated for the whole section and has been added to Fig. 1.14.

The load capacity of the member in this example will be considerably increased if it is possible to support the prestressed beam by propping before placing the added concrete. The only bending stress in the beam will be that caused by its own weight and the weight of the added concrete will be carried by the stiffer composite section. The calculation will then be as follows

Weight of prestressed beam alone

$$= 2.28 \text{ kN/m (155 lbf/ft)}.$$

Moment due to weight of beam

$$= 2.28 \times 8.5^2 \times 0.125 = 20.6 \text{ kN m (182 000 lbf in)}.$$

Resultant stress at soffit

$$= 0.8 \times 8.46 - 20.6 \times 10^6 \times \frac{140}{1250 \times 10^6} = 4.46 \text{ N/mm}^2 \text{ (645 lbf/in}^2\text{)}.$$

Permissible additional moment on composite member

$$= 4.46 \times 16.6 \times 10^6 \times 10^{-6} = 74.0 \text{ kN m (650 000 lbf.in)}.$$

Total permissible additional uniformly distributed load on composite member

$$= \frac{74.0}{8.5^2 \times 0.125} = 8.20 \text{ kN/m (550 lbf/ft)}.$$

Weight of added concrete

$$= 5.52 - 2.28 = 3.24 \text{ kN m (220 lbf/ft)}.$$

Permissible imposed load

$$= 8.20 - 3.24 = 4.96 \text{ kN/m (330 lbf/ft)}.$$

Deflexion

The deflexion of structural members is governed by the differential equation

$$\frac{\mathrm{d}^2 y}{\mathrm{d}x^2} = \frac{M}{EI}.$$

For concrete members with linear deformation the flexural rigidity EI is independent of M and may be obtained by calculating I for the equivalent concrete section and using the appropriate value of E for the concrete.

A convenient method of calculation is afforded by Mohr's two moment-area theorems. The notation and sign convention are defined in Fig. 1.15, with upward deflection (y) and moment causing sagging curvature considered positive.

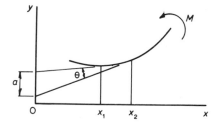

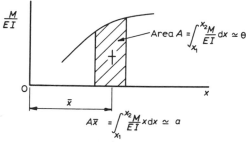

Figure 1.15   Mohr's moment-area theorems

1st theorem

Integrating the above differential equation between the horizontal limits x_1 and x_2

$$\left[\frac{dy}{dx}\right]_{x_1}^{x_2} = \int_{x_1}^{x_2} \frac{M}{EI}\,dx.$$

The physical interpretation of this equation is that the difference of slope of the member between x_1 and x_2 is equal to the area of the diagram M/EI by x between the vertical intercepts at x_1 and x_2.

2nd theorem

Multiplying each side of the differential equation by x and integrating by parts

$$x\frac{d^2 y_2}{dx^2} = \frac{M}{EI} \cdot x$$

$$\left[x\frac{dy}{dx} - y\right]_{x_1}^{x_2} = \int_{x_1}^{x_2} \frac{M}{EI} x\,dx.$$

Since $y - x\,dy/dx$ is the ordinate at $x = 0$ of the tangent to the deflected profile at the point x, the left hand side of the above equation represents the vertical distance between the tangents at x_1 and x_2 at $x = 0$. This is seen to be equal to the moment about the origin of the area of the M/EI diagram between the lines $x = x_1$ and $x = x_2$.

The deflexion of a symmetrically loaded simply supported beam at the midspan point is readily obtained from the second moment-area theorem since the tangent is horizontal at this point. The deflexion is therefore the moment of the area of one half of the M/EI diagram about the support. More complicated problems may be solved by the combined use of the two moment-area theorems.

When a member is prestressed it undergoes bending deflexion under the moment Pe resulting from the eccentricity e of the prestressing force P, and this may be calculated by the same method. To maintain the sign convention stated above, the sign of e will be positive when P acts above the centroid of the section and negative when it is below the centroid. If the member is prismatic, with the prestressing force constant along its length, the M/EI diagram will therefore be represented by the trajectory of the prestressing force, measuring the eccentricity from the axis of the member and multiplying it by P/EI. For strict accuracy I should be the value for the section used in the calculation of prestress (Fig. 1.11(a)).

A simple example is that of a simply supported beam of span l with straight pre-tensioned wires with the prestressing force acting a distance e below the centroid. The beam will be subject to a moment of value $-Pe$ constant along its length and from the second moment-area theorem the upward deflexion of the end, measured from the horizontal tangent at the midspan point, will be

$$-\frac{Pe}{EI} \cdot \frac{l}{2} \cdot \frac{l}{4} = -\frac{Pel^2}{8EI}$$

(the negative sign shows the deflexion to be downward). The centre of the beam will therefore deflect upward by this amount.

More complicated moment diagrams or prestressing force trajectories can be handled by dividing their areas into simple figures of rectangular, triangular or parabolic shape. Fig. 1.16 gives the necessary areas and positions of the centroids of these figures, and the following is an example of the application of the method.

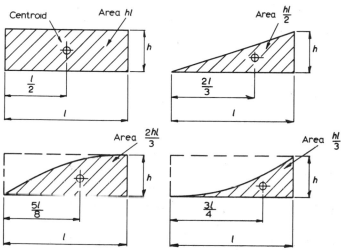

Figure 1.16   Basic shapes for computation of moment areas

EXAMPLE

Figure 1.17 shows the elevation and cable profile of a simply supported prismatic prestressed concrete beam with cantilevers at each end. The figures in square brackets indicate the vertical distance of the cable above or below the centroid of the beam section, the second moment of area of which is 3120×10^6 mm^4 (7500 in^4). If the prestressing force is 980 kN (220 000 lbf) and the modulus of elasticity of the concrete is 27.5 kN/mm^2 (4×10^6 lbf/in^2) calculate the deflexion of the ends of the cantilever above or below the level of the supports

(a) due to the prestress alone

(b) due to the combined effect of the prestress and a total uniformly distributed load of 11.6 kN/m (800 lbf/ft).

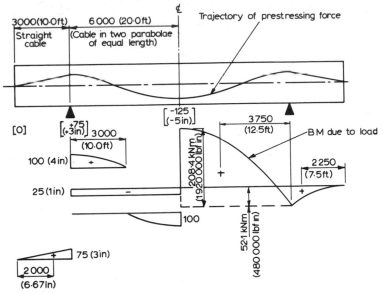

Figure 1.17   Example of calculation of deflexion

The deflexion of the end of the cantilever may be calculated in two possible ways:

(1) By calculating the slope of the beams at the supports by the first moment area theorem and obtaining the deflexion of the end measured from the tangent at the support by the second theorem.

(2) By using the second theorem to calculate the deflexion first of the support and then of the end relative to the midspan point where the tangent is horizontal.

The first method will be used for (a) and the second method illustrated by the solution of (b).

(a) the trajectory diagram of the prestressing force is analysed into its component areas in Fig. 1.17. Since $EI = 27.5 \times 3120 \times 10^6 = 86\,000 \times 10^6$ kN mm^2 (30×10^9 lbf in^2) increase of slope between left hand support and midspan

$$= \frac{+980}{86\,000 \times 10^6} \left[\frac{2}{3} \times 100 \times 3.0 - 25 \times 6.0 - \frac{2}{3} \times 100 \times 3.0 \right] \times 10^{-3} = -0.001\,71.$$

Since the slope is zero at midspan, rotation of beam at left hand support is anti-clockwise.

Deflexion of end from tangent at support

$$= \frac{+980}{86\,000 \times 10^6} \left[\frac{1}{2} \times 75 \times 3.0 \times 2.0 \right] \times 10^{-6} = +2.57 \text{ mm } (0.106 \text{ in}).$$

Deflexion of end of cantilever from support

$$= -3000 \times 0.001\ 71 + 2.56 = -2.57 \text{ mm } (-0.106 \text{ in.})$$

i.e. a downward deflexion.

(b) The bending moment diagram for a total load of 11.6 kN/m (800 lbf/ft) is shown in Fig. 1.17, divided into three component areas.

Deflexion of supports relative to midspan

$$= \frac{1}{86\ 000 \times 10^6} \left[\frac{2}{3} \times 208.4 \times 6.0 \times 3.75 - 52.1 \times 6.0 \times 3.0 \right] \times 10^9$$

$$= 25.5 \text{ mm } (1.07 \text{ in}).$$

Deflexion of ends relative to midspan

$$= \frac{1}{86\ 000 \times 10^6} \left[\frac{2}{3} \times 208.4 \times 6.0 \times 6.75 - 52.2 \times 6.0 \times 6.0 - \frac{1}{3} \times 52.1 \times 3.0 \right.$$
$$\left. \times 2.25 \right] \times 10^9 = 42.5 \text{ mm } (1.78 \text{ in}).$$

Deflexion of ends above supports

$$= 42.5 - 25.5 = 17.0 \text{ mm } (0.71 \text{ in}).$$

Resultant deflexion of ends due to prestress and load

$$= 17.0 - 2.6 = 14.4 \text{ mm } (0.60 \text{ in}) \text{ upwards.}$$

Shear and principal stresses

The shearing stress v at a point in a member in which the bending stress distribution can be assumed to be linear, is given by

$$v = \frac{V}{Ib}$$

where

V = shearing force,

S = statical moment ($S = A'\bar{y}$ as in Fig. 1.18),

I = second moment of area of section about its centroid,

b = breadth of section at given point.

The distribution of the shearing stress, calculated from the above expression, is illustrated for a rectangular section and an I section in Fig. 1.18. In the former

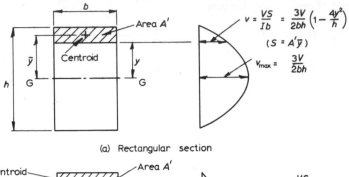

$$v = \frac{VS}{Ib} = \frac{3V}{2bh}\left(1 - \frac{4y^2}{h}\right)$$

$$(S = A'\bar{y})$$

$$V_{max} = \frac{3V}{2bh}$$

(a) Rectangular section

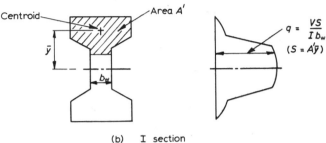

$$q = \frac{VS}{Ib_w}$$

$$(S = A'\bar{y})$$

(b) I section

Figure 1.18   Calculation of shearing stress

the distribution is parabolic while in the latter the high shear stress occurs in the thin web.

The shearing stress in uncracked members in which the deformation is assumed to be linear is thus obtained using the values of S and I for the total equivalent concrete section, In a prestressed member, however, when the line of the prestressing force is inclined to the axis at an angle θ an additional shearing force of magnitude $P \sin \theta$ will be introduced. Normally this will be of opposite sign to V, as for example, when cables are bent upwards at the end of a beam, and will therefore reduce the shearing stress.

At each point the shearing stress is normal to the axis with an equal complementary shearing stress in the axial direction. Where there are also direct stresses f_x and f_y in the axial and normal directions there are two mutually perpendicular planes on which the resultant is a maximum and minimum normal stress with no shearing stress (Fig. 1.19): these are the planes of maximum and minimum principal stress. The value of the principal stresses f_{max} and f_{min}

$$f_{\substack{max \\ min}} = \frac{f_x + f_y}{2} \pm \sqrt{\left(\frac{f_x - f_y}{2}\right)^2 + v^2}$$

and the angle of inclination by

$$\tan 2\phi = \frac{2v}{f_x - f_y}.$$

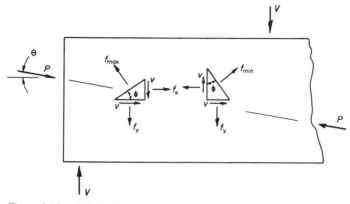

Figure 1.19   Principal stresses in a prestressed member

The Mohr's circle construction is useful if a graphical interpretation is desired.

If one considers the stress condition at the centroidal axis of a member without prestress, the direct stresses will be zero ($f_x = f_y = 0$) and the principal stresses will be a compressive and a tensile stress, each equal in value to the shearing stress v and inclined at 45° to the axis. Because of the weakness of concrete in tension the principal tensile stress may be sufficient to cause diagonal cracking, rather inaccurately termed 'shear cracking', in regions of high shear as is a common occurrence in reinforced concrete beams. In regions where the bending stress is high and the shearing stress low, the maximum principal stress will approach the value of the direct stress f_x due to flexure.

In a prestressed member the maximum shearing stress at the centroid of a section will be accompanied by a direct stress, f_x, equal to the average prestress. This will result in a reduction of the value of the principal tensile stress, thereby increasing the value of the load at which diagonal cracking will occur, while the angle between the principal tensile stress and the axis of the member will be greater than 45°. If it is also possible to apply a vertical prestress (f_y) both principal stresses may become compressive, but this involves the use of a number of short vertical or steeply inclined tendons and is only done where high shear presents a particular problem.

In the absence of vertical prestress, a more convenient arrangement of the principal stress formula for calculation is

$$f_{\substack{max \\ min}} = \frac{f_x}{2} \left\{ 1 \pm \sqrt{1 + \left(\frac{v}{f_x}\right)^2} \right\}.$$

EXAMPLE

Figure 1.20 shows the cross section of a post tensioned prestressed concrete beam. The prestressing force is 1330 kN (300 000 lbf) and at the point indicated on the diagram it acts upwards at an inclination of 1 in 20.

If at this section the design moment is 461 kN m (4 000 000 lbf in.) and the shearing force 333 kN (75 000 lbf), calculate the magnitude and direction of the greatest principal tensile stress.

The area of the cross-section is 196 000 mm² (305 in²) and the second moment of area is 18 600 x 10⁶ mm⁴ (44 720 in⁴) while the horizontal component of the prestressing force is

$$P \cos \theta = 1330 \times \frac{20}{\sqrt{(20^2 + 1^2)}}$$

$$= 1320 \text{ kN (299 000 lbf).}$$

Compressive prestress at bottom

$$= \frac{1320 \times 10^3}{196\ 000} \left[1 + \frac{445 \times 295}{95\ 000} \right] = 16.00 \text{ N/mm}^2 \text{ (2320 lbf/in}^2\text{).}$$

Tensile prestress at top

$$= \frac{1320 \times 10^3}{196\ 000} \left[\frac{445 \times 295}{95\ 000} - 1 \right] = 2.55 \text{ N/mm}^2 \text{ (380 lbf/in}^2\text{).}$$

Bending stress at top and bottom

$$= 461 \times 10^6 \times \frac{445}{18\ 600 \times 10^6}$$

$$= 11.00 \text{ N/mm}^2 \text{ (1565 lbf/in}^2\text{).}$$

The diagram of prestress and bending stress is therefore as shown in Fig. 1.20.

Effective shearing force

$$= 333 - 1330 \times \frac{1}{\sqrt{401}} = 267 \text{ kN (60 050 lbf).}$$

The shearing stress may now be calculated and the distribution diagram has been added to Fig. 1.20.

It will be seen from the diagrams of resultant direct stress and shearing stress that the principal tensile stress could reach its maximum value anywhere between the centroid and the lower root of the web. The stress is therefore

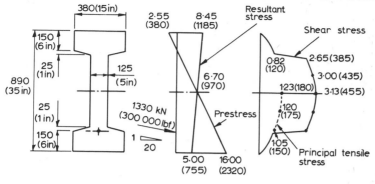

Stresses in N/mm²(lbf/in²)

Figure 1.20   Example of calculation of shear stress and principal tensile stress

investigated at three points in this region. Using the positive sign for tension the calculation for the principal stress at the centroid is

$$f_{\substack{max \\ min}} = \frac{-6.70}{2}\left\{1 \pm \sqrt{1 + \left(\frac{2 \times 3.13}{6.70}\right)^2}\right\}$$

$$= +1.23,\ -7.97\ \text{N/mm}^2\ (+180,\ -1150\ \text{lbf/in}^2).$$

$$\theta = \tan^{-1}\frac{2 \times 3.13}{-6.70} = +68°\ 20',\ -21°\ 40'.$$

The principal tensile stress of 1.23 N/mm² (180 lbf/in²) is found to be greater than the values calculated for the other two points and given in Fig. 1.20. In some instances the opposite may be true, but in order to shorten design calculations it is normal practice to check the principal tensile stress only at the centroid, making allowance for the possible small error by reducing the permissible value specified.

The principal compressive stress at design loads is almost always too small to be of direct significance, although in the F.I.P./C.E.B. proposals high values are considered to influence the principal tensile stress at which diagonal cracking occurs. The British Code, however, specifies only a permissible value of the principal tensile stress which is related to the strength of the concrete and is sufficiently conservative to take some account of shrinkage effects.

Shear and principal stresses due to torsion

Torsion, that is twisting about the axis of a member, sets up shearing stresses in the transverse and complementary planes which augment or diminish the shearing stresses associated with flexure. If the principal tensile stresses accompanying torsion become sufficiently great cracking occurs in a spiral direction around the member.

The torsional shearing stress at a point in a solid or hollow cylindrical shaft may be calculated from the well-known formula,

$$v = \frac{Tr}{J}$$

where T = torsional moment, r = radial distance of point from axis, J = second moment of area of transverse section about longitudinal axis ('Torsional moment of inertia').

The maximum shearing stress in a solid cylindrical shaft diameter h therefore occurs at the surface and is equal to $T/h^3\!\!/_{16}$.

The derivation of the above formula depends on the fact that the transerve sections remain plane after twisting of the shaft. This does not apply to members of non-circular cross section which undergo warping of the faces and the analysis of which is more involved. A number of approximate formulae have, however, been proposed to estimate the maximum torsional shearing stress in rectangular and flanged members.

In a solid member the most severe torsional shearing stresses occur at the centres of the sides and can be approximately obtained from a general formula of the type given in the German Code:

$$v = \frac{\psi T}{Ad}$$

where A = area of cross section

d = diameter of inscribed circle touching points of maximum stress (i.e. h for stress at centre of short sides of a rectangle and b for centre of long sides).

ψ is a coefficient which is equal to 4 for a circular section and usually lies between 3 and 5 for rectangular sections, for which the following two approximate formulae have been proposed.

$$\psi = 3 + \frac{2.6}{d/b + 0.45} \text{ (Bach)}$$

$$\psi = 3 + 1.8 \, \frac{b}{d} \qquad \text{(St. Venant)}$$

In the recent proposals of the American Concrete Institute a fixed value of $\psi = 3$ has been recommended. This is also considered satisfactory for box sections in which the wall thickness is not less than one quarter of the shorter overall dimension. For thinner walls, down to one tenth of the shorter side, the coefficient is increased to $3h_{min}/h_w$ where h_{min} is the dimension of the shorter side and h_w the thickness of the wall.

In a flanged section, an approximate expression for the shearing stress in a flange or web of thickness h_w is

$$v = \frac{3T h_w}{\Sigma(h_{min}{}^3 h_{max})}$$

where h_{min} and h_{max} are respectively the shorter and larger dimension of each component rectangle.

In prestressed members the total shearing stress, obtained by adding the stresses due to shearing force and torsion, may be used to calculate the principal stresses by the formula given above.

Shear in composite members

The importance of adequate bonding of the two or more parts forming a concrete member has already been noted, and this may be checked by calculating the shearing stress at the interface. The British Code specifies permissible values of the shearing stress for a prepared rough contact surface or for a smooth or rough surface with stirrups linking the two parts.

To calculate the principal tensile stress in a composite member the shearing stresses occurring in the prestressed part acting alone should be added to those resulting from load acting on the whole composite section, as was done for bending stresses. The principal tensile stress is then calculated for the total shearing stress and the total direct stress due to prestress and bending at the point required.

2

Cracked elements with linear deformation

Types of reinforcement

The design of normal reinforced concrete members over the last fifty years has assumed the presence of cracks wherever tensile stress occurs in the concrete, so that the required tensile resistance has to be provided by reinforcement, usually in the form of steel bars. Reinforcing bars of circular section are supplied in a range of preferred metric diameters from 6 mm to 40 mm with 5, 45 and 50 mm as additional sizes. Table 2.1 shows the main British sizes of bar with the nearest metric equivalent.

Until the Second World War mild steel was used for almost all reinforcement, but there has since been an increasing use of high strength steel, first cold-worked and more recently hot-rolled. The use of mild steel as specified in B.S. 785 is now generally restricted to reinforcement requiring sharper bends than the reduced ductility of high strength steel permits. The requirements for high strength steel, which is almost always deformed to improve the bond with the concrete, are specified in B.S. 875 for hot-rolled bars and in B.S. 1144 for cold-worked bars. Welded fabric of cold-drawn wire, as specified in B.S. 1221, has found widespread use in reinforcing slabs and panels.

Table 2.1   *Sizes of reinforcing bar*

Metric				(U.K.) (Nearest Equivalent)	
Dia.		Area		Dia.	Area
mm	in	mm²	in²	in	in²
5*	0.197	20	0.030	3/16	0.028
6	0.236	28	0.044	1/4	0.049
8	0.315	50	0.078	5/16	0.077
10	0.394	78	0.122	3/8	0.110
12	0.472	113	0.175	1/2	0.196
16	0.630	200	0.311	5/8	0.307
20	0.787	314	0.486	3/4	0.442
25	0.975	491	0.748	1	0.785
32	1.385	804	1.500	1 3/8	1.484
40	1.576	1255	1.95	1 1/2	1.767
45*	1.773	1590	2.47		
50*	1.970	1960	3.05	2	3.142

* Additional sizes

Behaviour of reinforcement in a tensile crack

The question of whether the presence of reinforcement affects the extensibility of concrete before actual cracking has been the subject of debate ever since Considère [2.1] put forward the hypothesis that the ultimate tensile strain of reinforced concrete was considerably greater than that of plain concrete. It is now generally believed that the ultimate tensile strain of concrete is about 100×10^{-6} and is independent of the presence or quantity of reinforcement, although the latter has a very marked effect on the subsequent increase of width of the cracks [2.2]. Recent work [2.3] has also shown how reinforcing bars act as crack arresters, inhibiting the propagation of cracks through the concrete.

It has been noted in the previous chapter that the tensile stress in reinforcement in an uncracked concrete member is relatively low. When the concrete fails in tension, however, there must be an increase in the stress in any reinforcement crossing the plane of failure, in order to maintain equilibrium. Thus in the simple condition of axial tension discussed in Chapter 1 (p. 3 and Fig. 1.2) the stress in the steel will increase to N/A_s when the concrete has failed in tension. In the example on p. 5 cracking of the concrete would increase the stress in the steel from about 19.6 N/mm² (2850 lbf/ in²) to 95 N/mm² (13 300 lbf/in²). The increased stress in the steel reinforcement is accompanied by an increased tensile strain relative to the strain in the adjacent concrete immediately before cracking. The factors affecting the consequent size of a crack will be discussed later, but it is immediately obvious that the width of a crack will be closely related to the stress in the steel. As it is desirable to restrict the width of cracks permitted in a concrete member the British Codes of Practice CP114 1957/1965 and CP116:1965 limited the tensile stress in reinforcement to 230 N/mm² (33 000 lbf/in²) regardless of the tensile strength of the steel.

Assumptions in the conventional theory of reinforced concrete

It will be recalled that in the previous chapter calculations for uncracked members were based on the assumptions that stress was proportional to strain, that the distribution of strain across any section was linear and that there was no slipping between the steel and adjacent concrete, so that the stresses in each material were related by the ratio of their moduli of elasticity. The first of these assumptions remains valid for the level of stresses under present consideration, but the other two must now be reconsidered.

Having regard to the discontinuity of strain in the concrete at a crack, and the fact that there must be some slipping of the reinforcement in the vicinity of a crack if the steel is not to undergo an infinite increase of strain and stress, both these assumptions can only be approximate. However, it can be demonstrated experimentally that the distribution of the *average* strains across a cracked reinforced concrete member in flexure is sensibly linear provided that the measurements are made over a gauge length sufficiently long to include several cracks. For the average strain measured adjacent to the reinforcement to be the same as the actual steel strain at the position of a crack, it would be necessary for there to be a complete failure of the bond and for the reinforcement between the cracks to slip. Clearly such extensive slipping is improbable in the early stages of cracking; nevertheless the error introduced by this assumption has been found to be fairly small, at the level of loading with which we are at present concerned.

The conventional theory whereby the calculations for cracked members are based on the above assumptions is now obsolescent, and is being replaced by an analysis directly related to the ultimate strength or other limit states. The conventional approach is still of considerable value, however, as the best available method for determining the stresses at the degree of loading to which reinforced concrete members are normally subjected. It may also be required in connection with calculations for the limit states of deflexion and cracking which often occur little beyond the normal loading.

Calculations for members subject to flexure

The calculation of the stresses in a cracked member with linear deformation proceeds from the conditions stated in the previous chapter, namely

1(a). Equilibrium of forces normal to the section.

(b). Equilibrium of internal and external moments at the section.

2. Compatibility of stress and strain in steel and concrete (assumed to be linear in transverse distribution).

These three conditions may be represented by three equations, the manipulation of which will be shown for the simplest example, that of a beam with tensile reinforcement (Fig. 2.1).

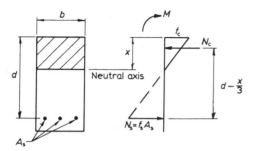

Figure 2.1   Bending stresses in cracked reinforced concrete member

$$N_s = N_c$$

(the tensile resistance of the concrete is neglected).

$$f_s A_s = \frac{f_c}{2} bx.$$

$$\frac{f_s}{f_c} = \frac{\xi}{2\rho}, \qquad\qquad 2.1.1(a).$$

where

$$\xi = x/d \quad \text{and} \quad \rho = A_s/bd.$$

$$M = N_s \left(d - \frac{x}{3}\right)$$

or

$$N_c \left(d - \frac{x}{3}\right)$$

$$= \rho\left(1 - \frac{\xi}{3}\right) bd^2 f_s$$

or

$$\frac{\xi}{2}\left(1 - \frac{\xi}{3}\right) bd^2 f_c. \qquad\qquad 2.1.1(b).$$

Since the strain in both the steel and the concrete are assumed to be proportional to the distance from the neutral axis

$$\frac{f_c/E_c}{x} = \frac{f_s/E_s}{d-x}.$$

$$\frac{f_s}{f_c} = \frac{\alpha(1-\xi)}{\xi} \qquad\qquad 2.1.2.$$

where

$$\alpha = \frac{E_s}{E_c}.$$

The stresses f_s and f_c are calculable from the two equations 2.1.1(b) if the depth of the neutral axis, x, is known; x may be obtained by equating the right hand sides of the remaining two equations, 2.1.1(a) and 2.1.2.

$$\frac{\xi}{2\rho} = \frac{\alpha(1-\xi)}{\xi},$$

$$\xi^2 + 2\rho\xi - 2\alpha\rho = 0,$$

$$\xi = -\alpha\rho + \sqrt{\alpha^2\rho^2 + 2\alpha\rho}.$$

It will be noted that the position of the neutral axis is independent of the level of loading (represented by M), and depends only on the modular ratio and reinforcement ratio.

EXAMPLE

A concrete floor slab 150 mm (6 in) thick is reinforced by 10 mm (⅜ in) diameter bottom bars at 120 mm (5 in) centres with 12 mm (½ in) cover of concrete. The slab has a span of 3 m (10 ft) and may be considered to be simply supported.

Calculate the maximum stress in the steel and concrete when the slab carries a uniformly distributed imposed load of 3800 N/m² (80 lbf/ft²) in addition to its own weight. If the permissible stresses are 140 N/m² (20 000 lbf/in²) in the steel and 7N/m² (1000 lbf/in²) in the concrete, what is the maximum imposed load on the slab? (Modular ratio = 15)

(University of Leeds, 1965)

Effective depth

$$d = 150 - 12 - \frac{10}{2} = 133 \text{ mm (5.32 in)}.$$

Area of steel per m width

$$A_s = 0.785 \times 10^2 \times \frac{1000}{120} = 655 \text{ mm}^2 \text{ (0.264 in}^2 \text{ per ft width)}.$$

Steel ratio

$$\rho = \frac{655}{1000 \times 133} = 0.00492 \text{ (0.00413)}.$$

$$\alpha\rho = 15 \times 0.00492 = 0.074 \text{ (0.062)}.$$

Neutral axis depth ratio

$$\xi = -0.074 + \sqrt{0.074^2 + 2 \times 0.074} = 0.318 \text{ (0.295)}.$$

Lever arm

$$z = d - \frac{x}{3} = 133 \left(1 - \frac{0.318}{3}\right) = 119 \text{ mm (4.80 in)}.$$

Self weight of slab

$$= 1.0 \times \frac{150}{1000} \times 2400 \times 9.81 = 3530 \text{ N/m}^2 \text{ (75 lbf/ft}^2\text{)}.$$

Total load

$$= 3530 + 3800 = 7330 \text{ N/m}^2 \text{ (155 lbf/ft}^2\text{)}.$$

Moment

$$M = 7330 \times (3.0)^2 \times 0.125 = 8250 \text{ N m/m width (23 250 lbf.in/ft width)}.$$

Tensile resistance of steel

$$N_s = \frac{M}{z} = \frac{8250}{119} \times 1000 = 69\,300 \text{ N/m width (4850 lbf/ft width)}.$$

Stress in steel

$$f_s = \frac{N_s}{A_s} = \frac{69\,300}{655} = 106 \text{ N/mm}^2 \text{ (18 400 lbf/in}^2\text{)}.$$

Stress in concrete

$$f_c = \frac{2N_s}{bx} = \frac{2 \times 69\,300}{1000 \times 0.318 \times 133} = 3.28 \text{ N/mm}^2 \text{ (515 lbf/in}^2\text{)}.$$

Note: this example was originally worked in Imperial units and owing to 'rounding-off' errors in the original data slightly different values are obtained for dimensionless quantities which are nominally the same.

The ratio of the stress in the steel to that in the concrete is $106/3.28 = 32.3$, and it is obvious from condition (2) that it is always the same for a given section. In this example the stress in the steel will therefore control the maximum permissible load, since when this stress attains its maximum permissible value of $140\ N/mm^2$ the stress in the concrete will only be

$$\frac{140}{32.3} = 4.34\ N/mm^2.$$

When

$$f_s = 140\ N/mm^2.$$

$$M = 140 \times 655 \times 119 \times 10^{-3} = 10\ 900\ N\ m/m\ \text{width (25 200 lbf in/ft width)}.$$

$$\text{Total load} = \frac{10\ 900}{3.0^2 \times 0.125} = 9700\ N/m^2\ (168\ lbf/ft^2)$$

Imposed load

$$= 9700 - 3530 = 6170\ N/m^2\ (93\ lbf/ft^2).$$

Other types of member

The foregoing three conditions also govern the calculations for more complicated types of reinforced concrete member, for example beams with compressive reinforcement, T and other non-rectangular sections and members subject to axial force and bending. In Table 2.2 the basic equations 1(a), 1(b) and 2 are given for different types of member, together with the formula derived for the solution of ξ.

Axial force and bending

The basic equations for a cracked reinforced concrete member subject to compression and bending may be written as in Table 2.2, with the axial force N appearing in equation 1(a) and the force N and its eccentricity e in 1(b). The working is less straightforward, however, and it is necessary to use all three equations to derive a cubic equation, the solution of which yields ξ. In view of this complication, direct calculation by this method has been rarely used in practice and it has been common to employ charts, usually in the form of load-moment diagrams (see Chapter 1 p. 8) for members with varying amounts of reinforcement.

The load-moment diagram may conveniently be plotted from the following expressions:

$$\frac{N}{bhf_c} = \frac{d}{h}\left\{\frac{\xi}{2} - \frac{\alpha\rho(1 - \xi)}{\xi} + (\alpha - 1)\rho'\frac{(\xi - d'/d)}{\xi}\right\},$$

(obtained from 1(a) and 2 in Table 2.2) and

$$\frac{N_e}{bh^2 f_c} = \frac{d^2}{h^2} \left\{ \frac{\xi}{2} \left(1 - \frac{\xi}{3} \right) + \frac{(\alpha - 1)\rho'(\xi - d'/d)(1 - d'/d)}{\xi} - \left(1 - \frac{h}{2d} \right) \frac{N}{bdf'_c} \right\}$$

(from 1(b) in Table 2.2).

These enable points on the curve, for a section with given reinforcement ratios ρ and ρ', to be plotted by inserting a series of values for ξ. An example is given in Fig. 2.2, showing the combined curve for the uncracked section, discussed in Chapter 1, and the cracked section.

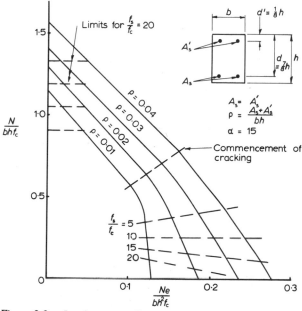

Figure 2.2   Load-moment diagram for cracked members with linear deformation

In British practice, up to and including Codes of Practice CP 114: 1957/1965 and CP 116: 1965, the conventional elastic theory was still permitted for design, as an alternative to the method discussed in Chapters 4 and 5. The condition imposed was that the calculated stresses in the steel and concrete should not exceed the permissible values specified. When using a load-moment diagram of the type shown in Fig. 2.2, the curves are curtailed by the limiting value of axial compression, as explained in Chapter 1. In addition it should be borne in mind that when the moment is large in relation to the axial load, the tensile stress f_s in the steel rather than the compressive stress f_c in the concrete may be the limiting criterion. This is readily seen from the dotted lines indicating the ratio of steel stress to concrete stress.

Table 2.2

	Rectangular section with tensile reinforcement	Rectangular section with tensile and compressive reinforcement
1 (a) Equilibrium of normal forces	$$\dfrac{f_s}{f_c} = \dfrac{\xi}{2\rho}$$	$$\dfrac{f_s}{f_c} = \dfrac{\xi + 2\rho'(\alpha - 1)\left(\xi - \dfrac{d'}{d}\right)}{2\,\rho\,\xi}$$
1 (b) Equilibrium of moments	$$M = \rho \left(1 - \dfrac{\xi}{3}\right) bd^2 f_s$$ or $$M = \dfrac{\xi}{2}\left(1 - \dfrac{\xi}{2}\right) bd^2 f_c$$	$$M = \left[\dfrac{(\alpha - 1)\,\rho'\left(\xi - \dfrac{d'}{d}\right)\left(1 - \dfrac{d'}{d}\right)}{\xi} + \dfrac{\xi}{2}\left(1 - \dfrac{\xi}{3}\right)\right] bd^2$$
2 Strain condition	$$\dfrac{f_s}{f} = \dfrac{\alpha(1 - \xi)}{\xi}$$	$$\dfrac{f_s}{f_c} = \dfrac{\alpha(1 - \xi)}{\xi}$$
Solution of ξ	$$\xi = -\alpha\rho + \sqrt{\alpha^2\rho^2 + 2\alpha\rho}$$	$$\xi = -(\alpha\rho + (\alpha - 1)\,\rho') + \sqrt{(\alpha\rho + (\alpha - 1)\,\rho')^2 + 2\left(\alpha\rho + (\alpha - 1)\,\rho'\dfrac{d'}{d}\right)}$$

Table 2.2

T section with tensile reinforcement	Rectangular section compression and bending
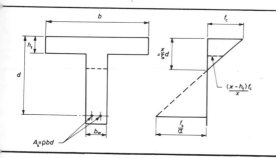	

$$\frac{f_s}{f_c} = \frac{\xi^2 - \left(1 - \frac{b_w}{b}\right)\left(\xi - \frac{d'}{d}\right)^*}{2\rho\xi}$$

$$N + \rho f_s bd = \left[\frac{\xi}{2} + \frac{(\alpha + 1)\rho'\left(\xi - \frac{d'}{d}\right)}{\xi}\right] bdf_c$$

$$\frac{\xi}{2}\left[1 - \frac{\xi}{3} - \right.$$

$$\left. \left(1 - \frac{b_w}{b}\right)\left(1 - \frac{h_f}{x}\right)^2\left(1 - \frac{\xi}{3} - \frac{2h_f}{d}\right)\right] bd^2 f_c$$

or neglecting b_w:

$$\rho\left[1 - \frac{h_f(3x - 2h_f)}{3d(2x - h_f)}\right] bd^2 f_s$$

$$\delta N d = \left[\frac{\xi}{2}\left(1 - \frac{\xi}{3}\right)\right.$$

$$\left. + \frac{(\alpha - 1)\rho'\left(\xi - \frac{d'}{d}\right)\left(1 - \frac{d'}{d}\right)}{\xi}\right] bd^2 f_c$$

where $\quad \delta = 1 + \frac{e}{d} + \frac{h}{2d}$

$$\frac{f_s}{f_c} = \frac{\alpha(1 - \xi)}{\xi}$$

$$\frac{f_s}{f_c} = \frac{\alpha(1 - \xi)}{\xi}$$

$$\left. - \frac{b_w}{b}\right) - \alpha\rho$$

$$\cdot \frac{\sqrt{\left(\left(1 - \frac{b_w}{b}\right)\frac{h_f}{d} + \alpha\rho\right)^2 + \frac{b_w}{b}\left[\left(1 - \frac{b_w}{b}\right)\frac{h_f^2}{d^2} + 2\alpha\rho\right]}}{b_w/b}$$

or neglecting b_w:

$$\xi = \frac{\left(\frac{h_f}{h}\right)^2 + 2\alpha\rho}{2\left(\frac{h_f}{d} + \alpha\rho\right)}$$

$$\frac{\xi^3}{6} + \frac{\xi^2}{2}(\delta - 1) + \xi\left(\delta\alpha\rho + \left(\delta - 1 + \frac{d'}{d}\right)(\alpha - 1)\rho'\right)$$

$$- \delta\alpha\rho - \frac{d'}{d}\left(\delta - 1 + \frac{d'}{d}\right)(\alpha - 1)\rho' = 0$$

where $\delta = 1 + \frac{e}{d} + \frac{h}{2d}$

---

n $\alpha < h_f$ equations are same as for rectangular section $b \times d$

The equivalent section method

The method described in the previous chapter can equally well be applied to the calculation of stresses in cracked members. The equivalent concrete area, moment of inertia and section moduli are calculated for a section from which the area of concrete in tension is omitted, and the standard bending formula used to calculate the stresses.

Before the dimensional properties of the cracked section can be calculated, however, it is necessary to establish the position of the neutral axis. For simple flexure the neutral axis passes through the centroid of the equivalent section, and the condition for equilibrium of the first moment of area about this axis yields the same equation for the depth of the neutral axis as that derived from the equilibrium and strain compatibility conditions, as in the example on p. 37. The equivalent section method is here illustrated for the example of a member with tensile and compressive reinforcement.

EXAMPLE

In the section shown in Fig. 2.3 the equilibrium condition of the first moment of area of the equivalent concrete section is

$$\frac{bx^2}{2} + (\alpha - 1)A_s'(x - d') = \alpha A_s(d - x),$$

and therefore

$$\xi^2 + 2\xi[\alpha\rho + (\alpha - 1)\rho' - 2\left[\alpha\rho + (\alpha - 1)\rho'\frac{d'}{d}\right] = 0$$

$$\xi = -[\alpha\rho + (\alpha - 1)\rho'] + \sqrt{\left\{[\alpha\rho + (\alpha - 1)\rho']^2 + 2\left[\alpha\rho + (\alpha - 1)\rho'\frac{d'}{d}\right]\right\}}.$$

The second moment of area of the equivalent concrete section about the centroidal axis is

$$I = \frac{bx^3}{3} + (\alpha - 1)A_s'(x - d')^2 + \alpha A_s(d - x)^2.$$

If f_c is the maximum compressive stress in the concrete the moment of resistance is given by

$$M = \frac{If_c}{x} = \left\{\frac{bx^3}{3} + (\alpha - 1)A_s'(x - d')^2 + \alpha A_s(d - x)^2\right\}\frac{f_c}{x}.$$

Putting

$$\alpha A_s(d - x) = \frac{bx^2}{2} + (\alpha - 1)A_s'(x - d')$$

and rearranging,

$$M = \left\{ \frac{\xi}{2} \left(1 - \frac{\xi}{3} \right) + \frac{(\alpha - 1)\rho'(\xi - d'/d)(1 - d'/d)}{\xi} \right\} bd^2 f_c.$$

This, and the above expression for ξ are identical to those obtained directly from the equilibrium and strain conditions 1(a), 1(b) and 2, (p. 37 and Table 2.2).

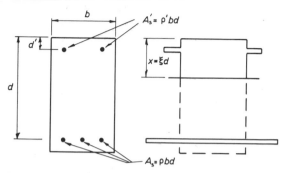

Figure 2.3   Equivalent concrete section for cracked member with compressive reinforcement

The equivalent section method is well suited for calculating the stresses in cracked composite members of structural steel sections with added concrete.

Cracking of prestressed members

Prestressed concrete members have in the past been designed to remain uncracked throughout their normal working life, although provision has been made for the eventuality of temporary flexural cracking under a rarely occurring excessive load. With the increase of knowledge and experience, however, the possibility is now envisaged of prestressed members in which cracks are permanent, subject to appropriate safeguards. The design issues involved in this development are discussed in Chapter 6, but it is appropriate at this point to examine the experimental and theoretical behaviour of a prestressed member after cracking.

If a prestressed concrete beam is sufficiently loaded, tensile stresses will be developed in the lower part, and when these become greater than the tensile strength of the concrete, cracking will occur. In laboratory tests cracks are usually observed at flexural tensile stresses between 3.5 and 6.5 N/mm² (500 and 900 lbf/in²), the higher values being associated with higher values of the prestress or with the presence of well-bonded and distributed steel close to the tensile face (e.g. in pre-tensioned beams). Experimental evidence indicates that under all conditions cracks of miscroscopic width are formed at a tensile stress of about 3.0 N/mm² (400 lbf/in²) but these are invisible to the naked eye until the above-quoted higher values of stress are attained.

The results of a typical test of a prestressed concrete beam are shown in Fig. 2.4. The load-deflexion curve is approximately linear until the tensile stress in the concrete at the soffit reaches a value of about 6 N/mm² (850 lbf/in²).

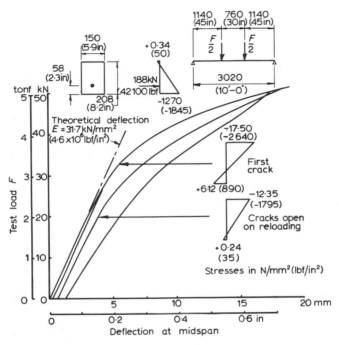

Figure 2.4   Load-deflexion diagram for cracked prestressed concrete beam

Beyond this point, which represents the appearance of visible cracks, the reduced stiffness of the beam results in a greater rate of increase of deflexion. The behaviour on unloading after cracking is of particular interest, for the cracks are found to close as the compressive effect of the prestress returns, so that they are no longer visible and the residual deflexion is very small. When the beam is again loaded the cracks remain closed as long as the surrounding concrete is in compression and reappear at approximately the decompression load as shown in Fig. 2.4. A prestressed member thus possesses a high degree of resilience even when loaded beyond the point at which the concrete cracks.

It was demonstrated in the first chapter that there is only a small increase of stress in the tensioned steel due to bending before cracking. After cracking, however, the stress will increase at a more rapid rate since the steel must then act as reinforcement, thereby providing the resistance which was previously supplied by the area of concrete which is now cracked. Provided that the stress in the steel is not too great, the strain will be proportional to stress, and the conventional methods of analysis used for reinforced concrete may be adapted to calculate the stresses in a prestressed member in the stage immediately after cracking.

This is, in fact, a special case of a member subject to axial compression and bending. The combined effect of the prestressing force P and the applied moment M is shown in Fig. 2.5, and the depth of the neutral axis x will

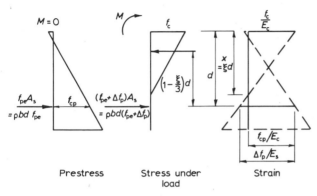

Figure 2.5   Stress and strain distribution in prestressed beam after cracking

obviously decrease as M increases. Neglecting the tensile resistance of the concrete, the two equilibrium equations are therefore

$$\rho b d(f_{pe} + \Delta f_p) = \frac{\xi}{2} b d f_c \qquad 2.2.1(a)$$

$$M = \frac{\xi}{2} \left(1 - \frac{\xi}{3}\right) b d^2 f_c \qquad 2.2.1(b)$$

where f_{pe} is the prestress in the steel and Δf_p the increase of stress due to the moment M.

If f_{cp} is the prestress in the concrete at the level of the prestressed steel, the tensile strain in the section at this level

$$= \frac{\Delta f_p}{E_s} - \frac{f_{cp}}{E_c}$$

The strain equation is therefore

$$\frac{\Delta f_p - \alpha f_{cp}}{f_c} = \frac{\alpha(1 - \xi)}{\xi} \qquad 2.2.2$$

These three equations yield the following expression for the position of the neutral axis:

$$\xi^3 \rho(f_{pe} + \alpha f_{cp}) + 3\xi^2 \left[\frac{M}{bd^2} - \rho(f_{pe} + \alpha f_{cp})\right] + 6\xi\alpha\rho\,\frac{M}{bd^2} - 6\alpha\rho\,\frac{M}{bd^2} = 0.$$

A graphical solution is given in Fig. 2.6 for a low and a fairly high value of the prestressed steel ratio, ρ, indicating the way in which the neutral axis rises as the moment increases.

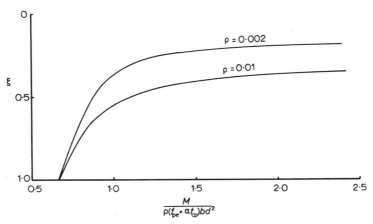

Figure 2.6   Position of neutral axis of cracked prestressed member

The stress f_p in the steel may now be obtained from the expression

$$f_p = f_{pe} + \Delta f_p = \frac{M}{A_s\left(d - \dfrac{x}{3}\right)}.$$

Hence

$$\frac{f_p}{f_{pe}} = \frac{M/bd^2 f_{pe}}{\rho\left(1 - \dfrac{\xi}{3}\right)}.$$

Figure 2.7 shows the relationship of steel stress to moment, the increase in the former being more rapid when the steel ratio is low. The maximum compressive stress, f_c, in the concrete is given by

$$f_c = \frac{2M/bd^2}{\xi\left(1 - \dfrac{\xi}{3}\right)}.$$

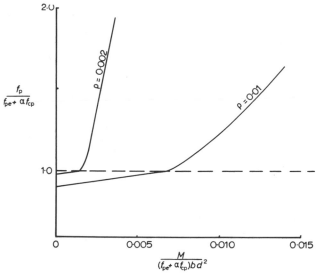

Figure 2.7   Theoretical stress in tensioned steel after cracking of prestressed member

Cracked composite prestressed members

A particular problem arises in the slab type of composite member, mentioned on p. 20 in that there may be a higher tensile stress in the added concrete on the tensile side of the neutral axis than the resultant tensile stress in the prestressed part of the member. This is demonstrated in the following simple example.

Figure 2.8 shows the cross section of a composite prestressed member formed by adding concrete above a prestressed plank. The prestress in the latter is 10 N/mm^2 (1450 lbf/in^2) uniformly distributed, and the plank is supported along its whole length until the added concrete has hardened.

If the concrete has the same modulus of elasticity in the two parts of the member, the section modulus for the bottom will be

$$\frac{300 \times 400^2}{6} = 8 \times 10^6 \text{ mm}^3 \text{ (488 in}^3\text{)}.$$

The resultant tensile stress at the soffit will therefore attain a value of 4 N/mm^2 (580 lbf/in^2), representing the approximate cracking stress, under a moment $M = (10 + 4) \times 8 \times 10^6 \times 10^{-6} = 112$ kN m (991 000 lbf in).

It is seen from Fig. 2.8, however, that at this load the tensile stress at the bottom of the added concrete is 7 N/mm^2 (1015 lbf/in^2). Cracking may therefore be expected to occur first at this point.

Tests, however, have revealed that cracks do not become visible under these conditions [2.4], even though the stresses are much higher than the values associated with cracking at the soffit of a member. Further research [2.2] has shown that cracking does in fact occur in the added concrete at about the normal values of stress and strain, but that the restraint of the uncracked prestressed part of the member results in a large number of cracks of such small width that they are only visible with the aid of a powerful microscope.

It may be argued that although such cracks are harmless in respect of corrosion, they will, by reducing the stiffness of the member, result in earlier appearance of cracks in the prestressed concrete at the soffit. This aspect of the problem may be examined in our example by using the conventional method of calculation for the partially cracked member, that is, by neglecting all tensile resistance of the added concrete, but assuming that the prestressed part of the section remains uncracked.

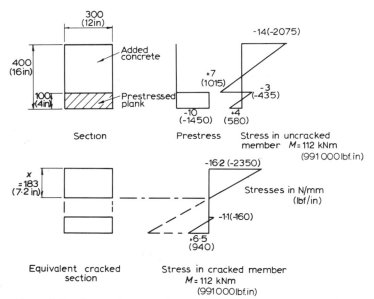

Figure 2.8   Stresses in composite prestressed member before and after cracking

The equivalent cracked section is illustrated in Fig. 2.8. From the first moments of area, the neutral axis depth is found to be 183 mm (7.2 in.).

The second moment of area about the neutral axis is therefore

$$I = \frac{300 \times 183^3}{3} + 300 \times 100 \left[\frac{100^2}{12} + (350 - 183)^2 \right] = 1478 \times 10^6 \text{ mm}^4$$
$$(3540 \text{ in}^4).$$

The section modulus for the bottom is

$$Z = \frac{1478 \times 10^6}{(400 - 183)} = 6.8 \times 10^6 \text{ mm}^3 \ (415 \text{ in}^3),$$

and the resultant tensile stress at the bottom is

$$\frac{112}{6.8} - 10 = 6.5 \text{ N/mm}^2 \ (940 \text{ lbf/in}^2).$$

The effect of cracking of the added concrete in this example is therefore either to increase the tensile stress at the soffit by 2.5 N/mm² (360 lbf/in²), or to reduce the load at which the soffit cracks (assuming this to take place at a stress of 4 N/mm² (580 lbf/in²)) by about 15 per cent.

The British Code permits the tensile stresses in the *in situ* concrete of composite prestressed members at the contact surface to be about 50 per cent greater than the stresses at the soffit of a member with pre-tensioned steel, and allows a further increase of 50 per cent provided the permissible tensile stress in the prestressed concrete unit is reduced by the same amount.

Conventional distribution of shearing stress in reinforced concrete members

The theoretical shearing stress related to the direct stress in a cracked section is derived by considering an element of the member bounded by two normal sections separated by a distance δx (Fig. 2.9). Let this element be intersected by a plane parallel to and distant y from the longitudinal axis of the member and at right angles to the plane of bending, represented by $X - X$ in the diagram. The forces acting horizontally on the part of the element above this plane arise from the compressive bending stresses on each side and the shearing stress v on the plane. If the maximum compressive stresses in the concrete on the two sides of the element are f_c and $f_c + \delta f_c$ respectively, the equilibrium condition is

$$v \, b \, \delta x = \delta f_c \left[\frac{bc}{2} - \frac{by^2}{2c} \right]^*,$$

$$v = \frac{\delta f_c}{2c\delta x} (c^2 - y^2).$$

Since

$$\frac{bcf_c z}{2} = M$$

where

$$z = d - \frac{c}{3},$$

$$f_c = \frac{2M}{zbc}$$

---

* In this example the American symbol c has been used to avoid confusion with the co-ordinate x.

and

$$\delta f_c = \frac{2\delta M}{zbc},$$

therefore

$$v = -\frac{1}{2c\delta x} \cdot \frac{2\delta M}{zbc}(c^2 - y^2)$$

$$= \frac{\delta M}{\delta x}\left(1 - \frac{y^2}{c^2}\right)\frac{1}{zb}.$$

When $\delta x \to 0$, since

$$\frac{\mathrm{d}M}{\mathrm{d}x} = V,$$

$$v = \frac{V}{zb}\left(1 - \frac{y^2}{c^2}\right).$$

The shear stress distribution above the neutral axis is therefore parabolic, with a maximum value V/zb at the neutral axis.

Below the neutral axis equilibrium conditions demand that the shear stress should be constant at the above value; it is, however, obvious that this is no more than a nominal stress as there is assumed to be a crack over this part of the

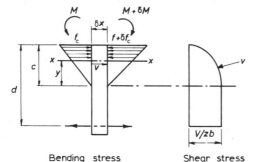

Bending stress   Shear stress
Figure 2.9   Shear stress distribution in a cracked section

section. Moreover, where the shearing force is accompanied by only a small moment, as in simply supported beams, it would be more correct to treat the member as uncracked, as in the previous chapter, so that

$$v \simeq \frac{3V}{2bh}, \quad \text{i.e.} \quad z \simeq \frac{2h}{3}.$$

Nevertheless, the shear stress as calculated by the former expression is widely used as an index of the severity of the shear effect on a reinforced concrete member.

Stress in shear reinforcement

The classical theory of shear reinforcement is due to Mörsch, and is here presented in a general form applicable to both prestressed and reinforced concrete. It is based on the fact that shear stress is associated with principal stresses, obliquely inclined to the axis of the member, and that the principal tensile stress will tend to cause diagonal cracking, often rather inaccurately termed 'shear' cracking. In the Mörsch theory the force obtained by integrating the principal tensile stress is assumed to be resisted entirely by the shear reinforcement.

In Fig. 2.10 the force due to the principal tensile stress, $f_{c \, max}$, inclined at ϕ to the cross section, acting over a length δx of the member is balanced by an element δN_{sv} of force in the shear reinforcement, inclined at α to the axis of the member.

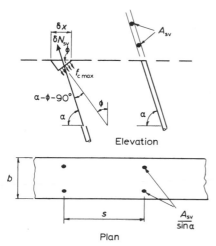

Elevation

Plan

Figure 2.10   Stress in shear reinforcement

The equilibrium condition for the forces in the direction of the principal tensile stress is therefore

$$\delta N_{sv} \cos(\alpha + \phi - 90°) = b \delta x \cos \phi \, f_{c \, max}.$$

Hence, over a length of member bounded by $x = x_1$ and $x = x_2$

$$N_{sv} = \int_{x_1}^{x_2} b f_{c \, max} \left(\frac{\cos \phi}{\cos(\alpha + \phi - 90°)} \right) dx$$

$$= b \int_{x_1}^{x_2} \left(\frac{f_{c \, max}}{\sin \alpha + \cos \alpha \tan \phi} \right) dx.$$

At the neutral axis of a reinforced concrete member

$$f_{c\,max} = v = \frac{V}{zb} \text{ and } \phi = 45°.$$

$$N_{sv} = \frac{1}{z\,(\sin\alpha + \cos\alpha)} \int_{x_1}^{x_2} V\,dx.$$

It will be noted that the integral is the area of the shear diagram.

When the shear is constant and the shear reinforcement consists of stirrups or bent-up bars of cross sectional area A_{sv} and spacing s, $N_{sv} = A_{sv}f_{sv}$ over the length s so that the stress in the shear reinforcement is given by

$$f_{sv} = \frac{f_{c\,max}}{\sin\alpha\,(\sin\alpha + \cos\alpha\tan\phi)}\left(\frac{bs\,\sin\alpha}{A_{sv}}\right)$$

$$= \frac{f_{c\,max}}{\sin\alpha\,(\sin\alpha + \cos\alpha\tan\phi)\rho_v},$$

where $\rho_v = \dfrac{A_{sv}}{bs\,\sin\alpha}$ i.e. the volumetric ratio, or the ratio in the horizontal section

(Fig. 2.10), of the shear reinforcement.

In a reinforced concrete member

$$f_{sv} = \frac{v}{\zeta\rho_v},$$

where $\zeta = \sin\alpha(\sin\alpha + \cos\alpha)$. ζ is equal to unity when $\alpha = 90°$ or $45°$ and has a maximum value of 1.207 when $\alpha = 67\frac{1}{2}°$.

The above formulae have been widely used for the dimensioning of shear reinforcement in the past and still form the basis of the more recent semi-empirical formulae discussed in Chapter 5.

Anchorage bond

It is obviously essential that the ends of bars which function as tensile reinforcement should be securely anchored. To meet this requirement the bond surface of the bar from the end to any point must be equal to the force in the bar at that point. Thus if f_b is the average bond stress (total force/surface area of bar) over a length l_b of bar of diameter d_b, and f_s is the stress required in the steel for equilibrium at the section at the end of this length, then

$$f_s\frac{\pi}{4}d_b^2 = f_b\,\pi d_b l_b,$$

$$f_b = \frac{d_b f_s}{4l_b}.$$

Although the above formula is widely used as a conventional criterion for design, by limiting the magnitude of f_b to permissible values, such as those specified in the British Code, which vary between 0.9 and 4.2 N/mm² (130 and 610 lbf/in²) depending on the type of bar and strength of the concrete, it must be appreciated that the bond stress is not uniformly distributed along the length of the bar, either in a reinforced concrete member or in the customary test in which a bar is pulled out from a block of concrete. This is due to the fact that the surface resistance to slipping is largely frictional and therefore dependent on the radial pressure between the bar and its surrounding concrete. The radial pressure is influenced by the lateral strain of the bar which in turn is related to the longitudinal strain by Poisson's ratio.

The distribution of bond stress f_b and direct stress f_s in the steel bar embedded in concrete, with a tensile force applied to one end, may be derived if it is assumed that the bond stress may be represented by the expression.

$$f_b = - \psi f_s + k.$$

The sign of the first term is negative since increasing tensile stress in the bar will cause transverse elastic contraction, thereby weakening the frictional component of the bond stress. The second constant k represents the 'adhesion' component which is assumed to be independent of the stress in the steel.

Considering the equilibrium of a length δx of the bar as in Fig. 2.11

$$\delta f_s \frac{\pi d_b^2}{4} = -(-\psi f_s + k) \pi d_b \, \delta x,$$

$$\frac{\frac{d_b}{4} \delta f_s}{-\psi f_s + k} = -\delta x,$$

$$\frac{-d_b}{4\psi} \log_e (-\psi f_s + k) + \text{constant} = -x.$$

If f_{so} is the tensile stress in the steel at the end of the bar, where $x = 0$, the value of the constant will be $\dfrac{d_b}{4\psi} \log_e (-\psi f_{so} + k)$, and therefore

$$\frac{-\psi f_{so} + k}{-\psi f_s + k} = e^{-4\psi x/d_b},$$

$$f_s = \frac{k}{\psi} - \left(\frac{k}{\psi} - f_{so}\right) e^{4\psi x/d_b},$$

$$f_b = -\psi f_s + k = (k - \psi f_{so}) e^{d_b x/4\psi}$$

Figure 2.11 illustrates the exponential increase of the bond stress with the distance from the tensioned end of the bar, while the stress in the steel diminishes. It will be noted that if f_{so} becomes sufficiently large, the bond resistance will vanish at the end $x = 0$, and slipping will commence. Even so the embedded length may be sufficient to take up the full force applied to the bar so that f_{so} may reach the tensile strength of the steel and the bar may fracture before a general bond failure occurs by slipping of the whole bar.

The conditions thus investigated do not correspond exactly to those in a reinforced concrete member in which the concrete surrounding the tensile reinforcement is itself cracked or in tension; moreover the actual relationship between steel stress and bond stress is undoubtedly more complex than the one used, particularly for deformed bars. However, laboratory tests have verified the general experimental form of the longitudinal distribution of stress [2.5], and bond tests which are more realistic than the simple 'pull-out' test have been devised to simulate actual structural conditions more closely.

Bars are frequently provided with U-shaped hooks or L-shaped 'bobs' at the ends to improve the anchorage and standard dimensions for these are specified in Codes of Practice. The British Code recommends the allowance of an additional bond length for each 45° of bend equal to the internal diameter of the bend, up to a maximum of 24 bar diameters. This may be compared with the maximum of 25 diameters now proposed by the European Concrete Committee, an allowance which is applicable to high strength ribbed bars bent to a diameter of curvature of 10 bar diameters; the allowance is reduced to 20 diameters for the more common hook bent to 5 bar diameters. Because of the difficulties of bending, hooks tend to be used less on high strength deformed bars: moreover the improved bond often makes them unnecessary.

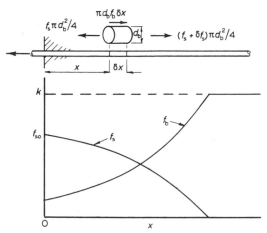

Figure 2.11   Distribution of bond stress

Tests have shown that the cover of concrete over the top and sides of a hook has a considerable effect on the bond resistance, as failure usually occurs by spalling of the concrete.

Flexural bond

The local bond stress associated with the shearing force, V, at a cracked section may be calculated by the conventional linear theory. The increase of moment δM over a length δx is given by

$$\delta M = f_b \Sigma u \, z \, \delta x,$$

where

$$f_b = \text{local bond stress,}$$

$$\Sigma u = \text{total perimeter of tensile reinforcement,}$$

$$z = \text{lever arm.}$$

Therefore, as $\delta x \to 0$

$$f_b = \frac{\dfrac{dM}{dx}}{\Sigma u \, z} = \frac{V}{\Sigma u \, z} \, .$$

The formula is used with specified permissible values of the local bond stress in reinforced concrete design according to the British Code, although it probably bears no more than a general relation to the actual bond stress which would appear to be distributed rather in the manner of Fig. 2.11, with a tendency to slipping at the cracks and with the steel stress a minimum midway between the cracks.

Deflexion of cracked reinforced concrete members

Simplified (unilinear) method

The equivalent (or transformed) cracked section is frequently used to calculate the deflexion of cracked members, by means of an ordinary linear bending calculation of the form

$$a = \frac{\beta l^2 M}{E_c I_c},$$

in which l is the span, M the moment at a given section and β a constant depending on the end-conditions, position of the given section and load distribution (e.g. $\beta = 5/48$ for the midspan section of a simply supported beam with a uniformly distributed load).

A simpler expression for the flexural rigidity may, however, be derived in terms of the equivalent steel section ($E_s I_s$) instead of the equivalent section ($E_c I_c$), given by

$$M = \frac{f_s I_s}{d - x} = f_s A_s z.$$

Therefore

$$I_s = A_s (d - x)z.$$

$$E_s I_s = E_s A_s (d - x)z \quad (= E_c I_c).$$

This method is permitted in the European Concrete Committee recommendations provided that the tensile reinforcement ratio (ρ) is not less than 0.005 for rectangular beams and 0.001 for T beams in which

$$\frac{b}{b_w} > 10.$$

Bilinear method

Deflexions calculated by this method tend to over-estimate those obtained experimentally. The reason for this is that the section on which the calculation is based occurs only at the cracks, the remaining part of the member being uncracked and therefore stiffer. The European Concrete Committee proposes a method utilising a bilinear deflexion curve (Fig. 2.12), which may be used as an alternative to the empirical method in the British Code. It has been observed from tests that the load-deflexion curve may be approximated by two straight lines, the slope of the first corresponding to the stiffness of the uncracked section and the slope of the second to that of the cracked section. The short term deflexion is therefore calculated by determining the cracking load and its corresponding deflexion based on the uncracked section, and adding to this the deflexion due to the portion of the load in excess of the cracking load, based on the cracked section. The expression proposed is

$$a = \frac{\beta l^2 M_r}{E_c I_c} + \frac{\beta l^2 (M - M_r)}{0.85 E_c I_r},$$

where I_c = second moment of area of uncracked equivalent concrete section,

I_r = second moment of area of cracked equivalent concrete section,

M_r = cracking moment at given section.

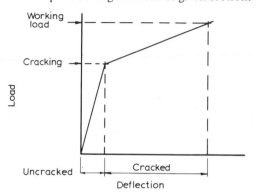

Figure 2.12   Bilinear method of calculating short-term
deflexion

EXAMPLE

Calculate the short-term deflexion of the reinforced concrete beam shown in Fig. 2.13 loaded at two symmetrical points. The cube strength of the concrete is 58 N/mm² (8400 lbf/in²) and the modulus of elasticity of the steel is 200 kN/mm² (29 x 10⁶ lbf/in²).

(a) Unilinear Method

$$A_s = 8 \times \frac{\pi}{4} \times 12.7^2 = 1015 \text{ mm}^2 \ (1.57 \text{ in}^2).$$

$$\rho = \frac{1015}{152 \times 271} = 0.0246.$$

$$E_c = 4.5 \sqrt{58} = 34.3 \text{ kN/mm}^2 \ (5 \times 10^6 \text{ lbf/in}^2).$$

$$\alpha = \frac{200}{34.3} = 5.8.$$

$$\alpha\rho = 5.8 \times 0.0246 = 0.142.$$

$$x = 271 \ (-0.142 + \sqrt{(0.142^2 + 2 + 0.142)}) = 111 \text{ mm} \ (4.35 \text{ in}).$$

$$z = 271 - \frac{111}{3} = 234 \text{ mm} \ (9.21 \text{ in}).$$

$$I_s = 1015 \ (271 - 111) \ 234 = 38 \times 10^6 \text{ mm}^4 \ (91.5 \text{ in}^4).$$

$$E_s I_s = 200 \times 38 \times 10^6 = 7600 \times 10^6 \text{ kN mm}^2 \ (2650 \times 10^6 \text{ lbf in}^2).$$

The deflexion of the beam at midspan is given by the formula

$$\frac{Fl_F \,(3l^2 - 4l_F^2)}{48 \, E_s I_s}$$

where l is the span and l_F the distance of each load $F/2$ from the nearest support. The calculated deflexion for a load of 50 kN (5.02 tonf) is therefore

$$a = \frac{50 \times 2130 \,(3 \times 6100^2 - 4 \times 2130^2)}{48 \times 7600 \times 10^6} = 27 \text{ mm (1.08 in)}.$$

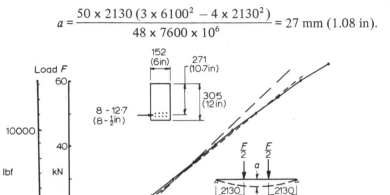

Figure 2.13   Calculated and experimental deflexion of reinforced concrete beam

The calculated deflexion line for this method is shown together with the experimental curve and can be seen to be conservative up to a load of about 30 kN (3 tonf).

(b) Bilinear method

It is first necessary to calculate the cracking moment M_r corresponding to the flexural tensile strength of the concrete, which according to the British Code is approximately 4.5 N/mm² (650 lbf/in²) for concrete of 58 N/mm² cube strength. The second moment of area of the uncracked section is 431×10^6 mm⁴ (1033 in⁴) and the centroid is 139 mm (5.5 in) from the bottom. Therefore

$$M_r = \frac{4.5 \times 431 \times 10^6}{139} \times \frac{1}{10^6} = 13.9 \text{ kN m (123 000 lbf in)}.$$

In the example there is a moment of 5.1 kN m at midspan due to 1.09 kN/m self weight of the beam, so that the applied moment required to produce cracking will be $(13.9 - 5.1) = 8.8$ kN m, corresponding to an applied load W of 8.3 kN (0.83 tonf).

For the given loading arrangement

$$\beta = \frac{aEI}{Ml^2} = \frac{1}{8} - \frac{1}{6}\left(\frac{l_F}{l}\right)^2 = 0.1046.$$

For the uncracked section

$$E_c I_c = 34.3 \times 431 \times 10^6 = 14\ 800 \times 10^6 \text{ kN mm}^2 \ (5170 \times 10^6 \text{ lbf in}^2).$$

The deflexion at a load of 50 kN (5.02 tonf) which corresponds to a moment of 53.2 kN m (471 000 lbf in) is therefore

$$a = \frac{0.104\ 6 \times 6100^2 \times 8.3 \times 10^3}{14\ 800 \times 10^6} + \frac{0.1046 \times 6100^2(53.2 - 8.3) \times 10^3}{0.85 \times 7600 \times 10^6}$$

$$= 29 \text{ mm } (1.14 \text{ in}).$$

The calculated bilinear deflexion curve gives a very good agreement with the experimental values in Fig. 2.13.

The deflexion of cracked prestressed members may be calculated by a similar method, using the increased cracking moment due to the prestress.

The method given in the American Code for calculating the deflexion of reinforced concrete members takes account of the bilinear character of the load-deflexion relationship by inserting a suitable 'effective' value of the flexural rigidity in the simple unilinear formula:

$$a = \frac{\beta l^2 M}{E_c I_c}.$$

The effective flexural rigidity $E_c I_c$ is obtained from the following expressions:

$E_c = D_c{}^{1.5} \times 33\sqrt{f_c'}$ in lbf/in^2 for concrete of density D_c between 90 and 155 lb/ft^3.

$$I_e = (M_r/M_{max})^3 I_c + [1 - (M_r/M_{max})^3] I_r,$$

where M_r = moment at cracking ($< M_{max}$),

$\quad I_c$ = second moment of area of uncracked concrete section,

$\quad I_r$ = second moment of area of cracked concrete section.

Width of cracks

The width of cracks in reinforced concrete (or cracked prestressed concrete) members has to be limited on account of the risk of corrosion of the reinforcement and the unsightly if not alarming appearance of large cracks, so that the width of a crack should not usually exceed about 0.3 mm (0.01 in). One of the most important variables affecting the width of cracks is the stress in the tensile reinforcement, and the design levels of stress permitted in past codes of practice have been chosen with a view to their relationship to crack width. With the need to make the most effective use of high strength steel, however, it is becoming increasingly necessary to make an independent calculation of crack width.

A number of semi-empirical formulae have been proposed, on a basis which is typified by the European Concrete Committee recommendations. The theoretical principle underlying these formulae is that the total force exerted by the concrete in tension midway between two adjacent cracks is transferred to the reinforcement by bond over a length equal to one half of the crack spacing. This force is equal to $f_{ct}A_c$ where

f_{ct} = tensile strength of concrete,

A_c = effective area of concrete in tension.

If f_b is the average bond stress, s the maximum spacing of the cracks, and Σu the total perimeter of the reinforcement, it will be seen from Fig. 2.14 that

$$f_b \frac{s}{2} \Sigma u = f_{ct}A_c,$$

$$s = \frac{2A_c f_{ct}}{\Sigma u f_b}.$$

Neglecting the relatively small tensile strain of the concrete the width of the crack, w, is equal to the total extension of the reinforcement over the length s. Therefore $w = s \times$ average strain.

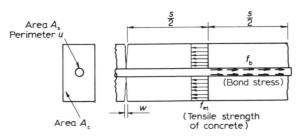

Figure 2.14   Cracking of reinforced concrete in tension

If the average strain is assumed to be $f(f_s)$, (a function of f_s which is the stress in the steel at the crack) and since $\Sigma u = 4A_s/d_b$ for reinforcement consisting of round bars of diameter d_b and total area A_s, then

$$w = \frac{d_b f_{ct} f(f_s)}{2 \rho f_b} \text{ where } \rho = \frac{A_s}{A_c} .$$

In the 1963 recommendations of the European Concrete Committee [2.6], the terms f_s, d_b and ρ were considered significant, but the formula was empirically modified so that the crack width was partly independent of ρ,

$$w = \left(4.5 + \frac{0.4}{\rho}\right) \frac{d_b f_s}{k} .$$

A different approach to the problem has more recently been made by a group of workers at the Cement and Concrete Association [2.7]. Starting from the assumption that no slip occurs at the interface between the steel and concrete, the theory of elasticity is used to show that the width of a crack increases with the distance from the bar. The formula originally proposed from a statistical study of experimental data was

$$w_{max} = \psi \frac{c f_s (h - x)}{E_s (d - x)} ,$$

where c = distance from point of measurement of crack to surface of nearest bar, ψ = a constant, of value 3.3 for deformed bars and 4.0 for plain round bars.

Clearly a certain degree of slipping must occur at the steel-concrete interface and the crack must therefore have some width at this point. Experience with composite members (p. 49) however, suggests that this width may well be extremely small, and that there may be a number of these cracks at the interface in addition to the one which propagates and becomes visible at the surface.

The above formula is supported by a large amount of experimental evidence and is important for two reasons; first that it indicates the great significance of the amount of concrete cover, and second that it reflects the relative unimportance of the bar diameter at normal levels of stress.

Appendix A of the British Code contains a modification of the above formula, applicable where the strain in the tensile reinforcement is limited to $0.8 f_y/E_s$.

$$\text{Design surface crack width} = \frac{3 a_{cr} \epsilon_m}{1 + 2 \left(\dfrac{a_{cr} - c_{min}}{h - x}\right)}$$

where c_{min} = minimum cover to the tension steel,

a_{cr} = distance from the point considered to the surface of the nearest longitudinal bar,

ϵ_m = average strain at the level where cracking is being considered. Allowing for the stiffening effect of the concrete between the cracks this may be obtained from the equation

$$\epsilon_m = \frac{\epsilon_1 \; 1.2 \, b_t \, h \, (a' - x)}{A_s(h - x)} \times 10^{-3},$$

where b_t = width of the section at the centroid of the tension steel,

a' = distance from compression face to point at which crack width is being calculated,

h = overall depth of the member,

ϵ_1 = strain at level considered ignoring stiffening effect of concrete,

x = depth of neutral axis found as used to determine ϵ_1

In assessing the strains it is suggested that the modulus of elasticity of the concrete should be taken as half the instantaneous value.

In the latest proposals of the European Concrete Committee, a modified formula has been introduced which takes account of the influence of cover:

$$w_{max} = \left[1.5c + \frac{\psi_1 d_b}{\rho} \right] \left[f_s - \frac{\psi_2 E_s}{\rho} \right] 10^{-5},$$

(linear dimensions in mm, stresses in N/mm^2)

where ρ = tensile reinforcement ratio, related to section defined in Table 2.3,

ψ_1 = constant related to tensile strength of concrete and bond (See Table 2.3),

ψ_2 = constant related to bond and loading conditions of member (See Table 2.3).

Width of cracks in prestressed members

If members are designed as partially prestressed ('Class 3'), so as to be cracked at normal levels of load it is particularly important to check the width of cracks. Less experimental data is at present available than for reinforced concrete members, and two somewhat different methods have been proposed.

The European Concrete Committee proposes a modified version of the formula for reinforced concrete quoted above.

Table 2.3 *Coefficients for crack width formula* (European Concrete Committee)

	Steel ratio	Reinforced Concrete		Prestressed Concrete	
	ρ	ψ_1	$\psi_2 E_s$ N/mm²	Min. ρ	ψE_s * N/mm²
Rectangular and T sections in simple bending	$\dfrac{A_s}{b_w d}$	0.04	0.75	0.010	0.37
Rectangular and T sections in compression and bending.	$\dfrac{A_s}{b_w(d-x)}$	0.07	1.20	0.016	0.60
Ties or beams with bottom flange	$\dfrac{A_s}{\text{(Total area)}}$ or $\dfrac{A_s}{\text{(Area of bottom flange)}}$	0.16	3.00	0.040	1.50

* For repeated loading $\psi E_s = 0$

$$w_{max} = \left(\Delta f_s - \frac{\psi E_s}{\rho}\right) 10^{-3},$$

(linear dimensions in mm, stresses in N/mm²)

where Δf_s = increase of stress in the steel with reference to the stress in the steel corresponding to decompression of the adjacent concrete,

ρ = steel ratio as defined in Table 2.3, but not to be less than the minimum values given,

ψ = constant related to bond and loading conditions of member (See Table 2.3).

In the British Code the calculated crack width is based on the strength of the concrete and on the so called 'hypothetical flexural tensile stress' in the concrete, that is, the stress calculated assuming the member to be uncracked. The proposed values of these tensile stresses are given in Table 2.4 in which allowance is made for the reduction of crack width by well distributed pre-tensioned tendons or additional reinforcement close to the tensile face of the concrete. 'Depth factors' are also recommended whereby the hypothetical stresses are progressively reduced by amounts of up to 30 per cent for members of greater depth than 400 mm.

Table 2.4   *Hypothetical flexural tensile stresses for Class 3 Structures* (British Standard Code of Practice)

	Limiting crack width mm	Stress for given strength of concrete N/mm²			Increase per 1 per cent of additional reinforcement*
		30	40	50 and over	
A Pre-tensioned tendons	0.1	–	4.1	4.8	4.0
	0.2	–	5.0	5.8	4.0
B Grouted post-tensioned	0.1	3.2	4.1	4.8	4.0
tendons	0.2	3.8	5.0	5.8	4.0
C Pre-tensioned tendons	0.1	–	5.3	6.3	3.0
distributed in the tensile	0.2		6.3	7.3	3.0
zone and positioned close					
to the tension force of the					
centre					

*Hypothetical tensile stress not to exceed $0.25 f_{cu}$.

REFERENCES

[2.1] Considère, A. (1898) *Genie civil,* **34**, 213, 229, 244, 268.

[2.2] Evans, R. H. and Kong, F. K. (June, 1964) The extensibility and microcracking of the in-situ concrete in composite prestressed concrete beams. *Struct. Engnr.,* **42**, 181-9.

[2.3] Romualdi, J. (June, 1964) Tensile strength of concrete affected by uniformly distributed and closely spaced short lengths of wire reinforcement. *Proc. Amer. Concr. Inst.,* **61**, 657-671.

[2.4] ———— (August, 1950) A test of a prestressed concrete railway bridge girder. Concr. *Constr. Engng,* **45**, 296-299.

[2.5] Glanville, W. H. (1930) Studies in reinforced concrete, I, Bond resistance. *Building Res.* Tech. Paper No. 10, H.M.S.O.

[2.6] *Recommendations for an international code of practice for reinforced concrete.* (1964) London. American Concrete Institute and Cement and Concrete Association.

[2.7] Base, G. D., Read, J. B., Beeby, A. W., and Taylor, H. P. J. (1966) *An investigation of the crack control characteristics of various types of bar in reinforced concrete beams. Research Report* No. 18, Part 1. London: Cement and Concrete Assocn.

3

Time-dependent deformation of elements under normal loading

Methods of calculation

Unlike many structural materials, concrete is not linear and elastic in its deformation, even at the normal level of design loading, but undergoes an additional deformation which is related to time. This time-dependent deformation may conveniently be divided into shrinkage, independent of the stress in the concrete, and creep, which is related to the magnitude of the stress. The extent of both shrinkage and creep depends on the composition of the concrete and on its environment, and although it is convenient to treat the two phenomena separately, they are to a certain degree inter-related.

The shrinkage and creep of concrete will obviously modify the stress and deformation of a structural member. Conventional calculations generally assume a linear stress-strain relationship, both for steel and concrete, and certain adjustments may be made to allow for the effects of shrinkage and creep while continuing to use this method. For example, it has already been explained in Chapter 1 how allowance is made for the effect of creep by reducing the value of the modulus of elasticity of the concrete. In a reinforced concrete member the effect of shrinkage is often to increase the tensile stresses in the concrete and may therefore be compensated by a reduction of the value of the calculated tensile stress permitted in design, as in B.S.C.P. 2007 for water-retaining structures. These methods, however, cannot be regarded as other than approximate, and there are occasions on which a more precise evaluation of the separate effects of shrinkage and creep is required.

A short account will be given of the principal factors governing the magnitude of the shrinkage and creep of concrete, followed by a description of a method of calculation for structural members which is simple in concept and sufficiently accurate in the majority of instances. The approximate method generally used for prestressed concrete will also be described.

Shrinkage of concrete

The main shrinkage of concrete is associated with the loss of uncombined water from the hardened cement paste. The sub-microscopic structure of hydrated Portland cement is now generally considered to be partly crystalline, but mainly a gel formed of thin crumpled plates of the order of 5000 Å in length and breadth but only about 30 Å in thickness [3.1]. The large surface forces exerted by these plate-like gel particles are responsible for the adsorption of some of the molecules of uncombined water, most of which is contained in larger pores in the paste. The loss of water from the larger pores by evaporation from the surface of the concrete reduces the vapour pressure so that some of the adsorbed water molecules are removed from the surfaces of the gel particles. The surface forces thus released cause the particles which are only about 15 Å apart to be drawn more closely together, resulting in a shrinkage of the whole mass. It is possible that a similar effect is produced by the increased surface tension force at the reduced meniscus where two gel particles meet at an angle.

Under controlled conditions the shrinkage of concrete which is usually expressed as a strain, occurs at a rate which diminishes with time and approaches a maximum, or limiting value. The shrinkage-time relationship can be approximately represented by an exponential equation of the form;

$$\epsilon = \epsilon_{cs\ max} (1 - e^{-kt}),$$

where ϵ_{cs} = shrinkage at time t, $\epsilon_{cs\ max}$ = limiting shrinkage at time $t = \infty$, and k is a constant, depending, like $\epsilon_{cs\ max}$ on the composition of the concrete and its environment.

In all but the most precise calculations the selected value of the shrinkage is treated as though it occurred instantaneously and the form of the shrinkage-time relationship is not taken into account. A rough guide to the rate of shrinkage occurring in practice is given in the British Code where it is suggested that one-half of the limiting shrinkage should be assumed to occur in the first month, and three-quarters in the first six months.

The temperature and relative humidity of the environment, by their influence on the evaporation of moisture from the concrete, have an important effect on the shrinkage, which is much greater for example in concrete members in a heated building than in members exposed in the open in temperate climates. Change of temperature and relative humidity may result in some recovery of shrinkage although the recovery is never observed to be complete. The permanent component of shrinkage has been attributed to the establishment of links by the surface forces as the gel particles are brought closer together. Owing to the varying rate and partial recovery of shrinkage the experimental shrinkage-time relationship of concrete under natural, as opposed to controlled laboratory conditions, is very irregular, as will be seen from Fig. 3.1.

Since moisture is lost most rapidly near the surface of a member, the shrinkage tends to be greatest in this region. There is thus a differential shrinkage effect which is most pronounced in members of large cross section, in which tensile stresses sufficient to cause cracking are frequently induced, even in plain concrete. A further result is that the overall shrinkage is less in a member of large cross section.

The magnitude of shrinkage has been found to vary considerably according to the composition of the concrete, the principal factors being as follows:

Water content [3.2]

This has a very important effect, and the amount of shrinkage is directly related to the water-cement ratio. An increase in the water content produces a correspondingly greater volume of gel, thereby increasing the potential shrinkage which will occur when moisture is lost.

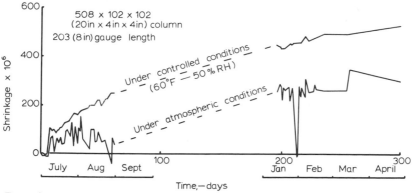

Figure 3.1   Shrinkage curves of concrete under controlled and atmospheric conditions

Aggregate content [3.3]

Particles of aggregate have a restraining effect on shrinkage of the cement paste which by itself is several times greater than that of concrete. The shrinkage of concrete is thus reduced by increasing the amount of aggregate, although this is achieved at the cost of greater internal stresses and microcracking between the aggregate and paste.

Type of aggregate [3.4]

It is obvious from the foregoing that the shrinkage will be greater when less rigid aggregates are used, and certain types, notably sandstones, are associated with undesirably high shrinkage, aggravated still further when the aggregates themselves undergo drying shrinkage.

Fineness of cement [3.5]

Pastes made with finer cements tend to shrink more, although the effect is less marked in concrete than in neat paste because of the restraint by the aggregate. The reduced shrinkage of paste made from coarse cement may itself be the result of restraint by the larger particles of unhydrated cement clinker.

Typical values of overall shrinkage have been recommended for structural calculations but should be used with caution in view of the many factors by which shrinkage may be affected. A limiting shrinkage of 300×10^{-6}, as given in the British Code for prestressed concrete members made by pre-tensioning, is commonly used. The code also contains more detailed data taking account of the cement and water content of the concrete, and includes correction factors for relative humidity and size of member.

Creep of concrete

The creep-time relationship of concrete at constant stress is of comparable form to the shrinkage-time relationship, and a maximum value, or limiting creep, may be approached asymptotically after several years. Some typical results of laboratory tests are given in Fig. 3.2. If the stress is reduced, recovery takes place also at a decreasing rate, approaching a final value. Part of the creep is permanent, and there is never complete recovery even after the removal of all the load.

The limiting creep, as also the intermediate values, is influenced by broadly the same factors of composition and environment as affect the shrinkage, together with two important additional parameters, namely the stress and the strength or degree of hydration of the concrete at which loading commences.

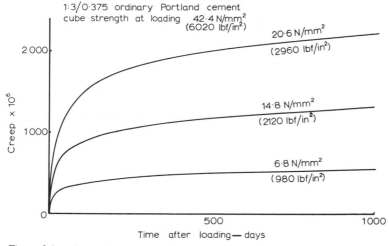

Figure 3.2   Creep-time relationship of concrete (Bennett and Loat)

It is seen clearly from Fig. 3.2 that creep is greater in the more highly stressed specimens. It is found, as in Fig. 3.3, to be directly proportional to the stress, up to a certain limit, which may be conservatively estimated as about one third of the cube strength; beyond this limit there is increased deformation, probably on account of internal cracking, particularly at the interfaces between the cement paste and the particles of aggregate. At lower stresses the creep per unit stress, or specific creep, may be assumed constant for a concrete of given proportions, except where loading commences at an early age.

The fundamental nature of the creep of concrete is not yet fully understood, but is believed to be a deformation of the gel structure of the hydrated cement accompanied by some redistribution of the gel water. It has been observed that the increase in the total deformation of a drying specimen is greater than can be accounted for by the increase in the shrinkage of an unloaded control specimen.

A number of exponential or hyperbolic expressions have been proposed to represent the creep-time relationship [3.6], and limiting values of the specific creep have been suggested. The British code recommends the figure of 36×10^{-6} per N/mm^2 (0.25×10^{-6} per lbf/in^2) for beams prestressed by post-tensioning at an age between 7 and 14 days and an increased value of 48×10^{-6} per N/mm^2 (0.33×10^{-6} per lbf/in^2), for pre-tensioning at an age between 3 and 5 days. These values are increased proportionately when the strength is less than $40 \ N/mm^2$ ($6000 \ lbf/in^2$). The above proposals are, however, based on the stress in the concrete at the time of prestressing, and since the subsequent loss of prestress will result in a lower rate of creep, they are not strictly comparable with the results of tests at constant stress.

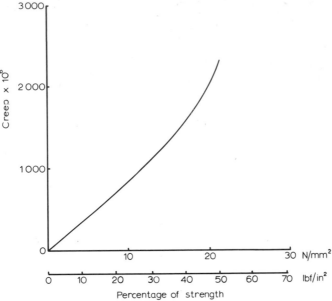

Figure 3.3   Relationship of creep to stress-strength ratio (Bennett and Loat)

Creep is sometimes expressed as a fraction of the elastic deformation, that is, as the product of the specific creep and the modulus of elasticity. The British Code also gives values for the creep coefficient ϕ thus defined, for varying values of cement content and water-cement ratio, with correcting factors for the effect of relative humidity and age of the concrete at the commencement of loading.

Method of calculation of stress and deformation

In structural members with reinforcement the concrete is not free to undergo its full amount of time-dependent deformation, but is restrained by the steel reinforcement, so that stresses are induced in both steel and concrete. The method of calculation to be described assumes that in the absence of such restraint the distribution of time-dependent strain would be linear across the member.

The procedure is based on the concept of imaginary forces which are considered to be applied to the reinforcement, and are of such magnitude as to cause a strain in the steel equal to the unrestrained time-dependent deformation of the adjacent concrete. At this first stage there is deformation, but the only stress is that in the reinforcement, produced by the imaginary forces.

Since there are actually no external forces on the member, the above imaginary forces must be cancelled by introducing equal and opposite external balancing forces. These must be considered to act on the member as a whole, and set up stresses in both the steel and the concrete which may be calculated by the usual method. The final stress in the concrete is therefore that due to the balancing forces, while in the steel the stress due to the imaginary forces must also be included. The final deformation of the member is the sum of the unrestrained time-dependent deformation in the first stage, and the deformation resulting from the balancing forces in the second stage.

The stresses thus calculated are added to the stresses due to loading, the method being applicable also to prestressed members. It will be illustrated by examples of the calculation of stresses due to shrinkage and creep. Where the use of formulae is preferred reference may be made to an earlier work [3.7].

Calculation of stresses due to shrinkage of concrete

The effect of a uniform shrinkage of unrestrained magnitude ϵ_{cs} on a beam with a single layer of reinforcement, is demonstrated in Fig. 3.4. In order to prevent restraint of the concrete, a compressive stress must be imparted to the steel of magnitude $\epsilon_{cs} E_{cs}$ requiring an imaginary force $\epsilon_{cs} E_s A_s$ applied to the

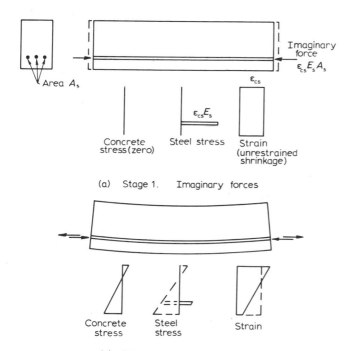

(a) Stage 1. Imaginary forces

(b) Stage 2. Balancing forces added

Figure 3.4 Calculation of stress and strain due to shrinkage of concrete

reinforcement parallel to the axis at each end of the beam (Fig. 3.4(a)). Since no external forces act on the beam, equal and opposite (i.e. tensile) forces have to be applied; these act on the reinforced member as a whole, resulting in a combination of tensile and bending stress in the concrete and a modification of the compressive stress already present in the steel (Fig. 3.4(b)).

EXAMPLE

Calculate the change of stress at the prestressed beam section shown in Fig. 3.5 resulting from a uniform shrinkage of 300×10^{-6}, given that the modulus of elasticity of the tensioned steel is 193 kN/mm^2 (28 $\times 10^6$ lbf/in^2) and of the untensioned steel 205 kN/mm^2 (30 $\times 10^6$ lbf/in^2).

The dimensional properties of the equivalent concrete section, assuming a modulus of elasticity of 32 kN/mm^2 (4.65 $\times 10^6$ lbf/in^2) are

$$A = 245\ 600 \text{ mm}^2 \ (380 \text{ in}^2),$$
$$y_{\text{inf}} = 486 \text{ mm (19.1 in)},$$
$$I = 31\ 000 \times 10^6 \text{ mm}^4 \ (74\ 500 \text{ in}^4).$$

The imaginary forces are

Tensioned steel:

$$3 \times 462 \times 300 \times 10^{-6} \times 193\ 000 = 80\ 400 \text{ N (18100 lbf)}.$$

Bottom untensioned steel:

$$4 \times 113 \times 300 \times 10^{-6} \times 205\ 000 = 27\ 800 \text{ N (6250 lbf)}.$$

Top untensioned steel:

$$2 \times 113 \times 300 \times 10^{-6} \times 205\ 000 = 13\ 900 \text{ N (3100 lbf)}.$$

The total balancing force is therefore 122 100 N (27 450 lbf), and the moment of this force about the centroid of the section is

$$80\ 400 \times 386 + 27\ 800 \times 456 - 13\ 900 \times 484 = 37\ 000\ 000 \text{ N mm (321 000 lbf in)}.$$

The differential stresses in the concrete and steel may now be calculated. For example, the stress in the concrete at the bottom will be

$$\frac{122\ 100}{245\ 600} + \frac{37\ 000\ 000 \times 486}{31\ 000 \times 10^6} = 0.50 + 0.57 = 1.07 \text{ N/mm}^2 \ (155 \text{ lbf/in}^2).$$

In the top cable the stress due to the imaginary force will be

$$-300 \times 10^{-6} \times 193\ 000 = -57.9 \text{ N/mm}^2 \ (8400 \text{ lbf/in}^2).$$

To this must be added the stress caused by the balancing force, that is, the stress in the concrete multiplied by the modular ratio

$$\left[1.07 - \frac{150}{1000}(1.07 + 0.10)\right]\frac{193\,000}{32\,000} = +5.4\ \text{N/mm}^2\ (780\ \text{lbf/in}^2).$$

The resultant differential stress in the steel is therefore

$$-57.9 + 5.4 = -52.5\ \text{N/mm}^2\ (7620\ \text{lbf/in}^2).$$

The complete shrinkage stress diagram is given in Fig. 3.5.

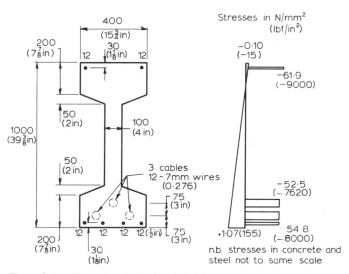

Figure 3.5 Example of calculated shrinkage stresses

Calculation of stresses due to creep of concrete

The method of imaginary and balancing forces may also be used to calculate creep stresses, assuming a linear distribution of strain due to unrestrained creep. In prestressed concrete calculations this is usually based on a value of specific creep applied to the stress in the concrete at transfer, although it would appear more rational to base the calculation of creep on the stress permanently maintained during the life of the member, that is, under the total dead load.

The procedure is shown diagrammatically in Fig. 3.6. Unlike the shrinkage calculation, in which a uniform unrestrained contraction is assumed, the imaginary forces will now be different in layers of steel at different levels; thus if the stresses in the concrete at the levels of the prestressed and non-prestressed steel are f_{cp} and f_{cs} respectively (Fig. 3.6(a)) the unrestrained creep at the two

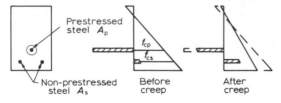

(a)   Resultant   stresses

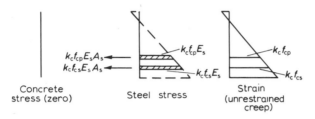

(b) Stage 1.   Imaginary forces

(c) Stage 2. Balancing forces added

Figure 3.6   Calculation of stress and strain due to creep of concrete

levels is obtained by multiplying these stresses by the specific creep k_c and the two imaginary forces are $k_c f_{cp} E_s A_p$, and $k_c f_{cs} E_s A_s$ (Fig. 3.6(b)). The effect of the equal and opposite balancing forces is then calculated as before.

EXAMPLE
If the initial stress in the tensioned steel in the previous example is 880 N/mm²
(128 000 lbf/in²) and the dead load moment is 200 kN m (1 735 000 lbf in),
calculate the change of stress in the steel and concrete due to a specific creep of
36 x 10⁻⁶ per N/mm² (0.25 x 10⁻⁶ per lbf/in²).

The initial stress in the concrete is first calculated as described in Chapter 1 and is shown in the left hand diagram of Fig. 3.7. This enables the required imaginary forces to be computed; for example, at the level of the upper

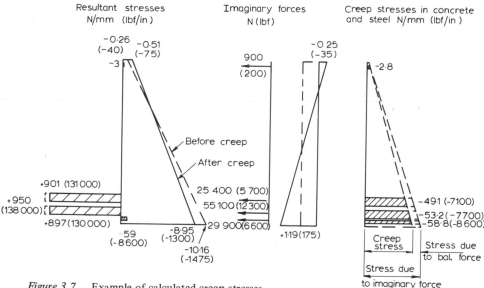

Resultant stresses
N/mm (lbf/in)

Imaginary forces
N (lbf)

Creep stresses in concrete
and steel N/mm (lbf/in)

Figure 3.7 Example of calculated creep stresses

tensioned cable the compressive stress in the concrete is 7.92 N/mm^2 (1150 lbf/in^2) and the imaginary force is therefore

$$7.92 \times 36 \times 10^{-6} \times 462 = 25\ 400\ \text{N (5700 lbf)}.$$

The sum of the imaginary forces in all the cables and reinforcement is 120 300 N and the moment of the forces about the centroid of the section is 44.45 kN m (385 000 lbf in). The change of stress in the concrete and steel is obtained as before and the results are given in the diagrams comprising Fig. 3.7.

Deformation due to shrinkage and creep

The deformation induced in a member by shrinkage or creep of the concrete consists of a combination of axial contraction and flexure. Since the strain corresponds to the stress in the steel (Figs. 3.4 and 3.6), it may readily be calculated.

In the present example the creep strain at the level of the bottom reinforcement is

$$-\frac{58.8}{205\ 000} = -287 \times 10^{-6}$$

and the strain at the level of the top reinforcement is

$$-\frac{2.8}{205\ 000} = -14 \times 10^{-6}$$

The strain at the centroid is therefore -154×10^6 and the curvature

$$\frac{1}{r} = \frac{287 - 14}{940} \times 10^{-6} = 0.290 \times 10^{-6} \; (\text{mm}^{-1}).$$

The axial contraction of a given length is calculated from the strain at the centroid while the deflexion is obtained from the standard formula putting $M/EI = 1/r$. For example, over a 20 m length with the given section the axial contraction will be

$$154 \times 10^{-6} \times 10 \times 10^3 = 3 \text{ mm}.$$

The upward deflexion will be

$$\frac{1}{r} \times \frac{L^2}{8} = 0.290 \times 10^{-6} \times \frac{20^2 \times 10^6}{8} = 14 \text{ mm}.$$

The deformation will of course be affected by variation in the height of the cables and changes in the reinforcement, when a numerical integration process will be necessary.

Complementary method

In some instances it is simpler to calculate the stress and deformation due to shrinkage and creep by a method which is the complement of the one already described.

The imaginary forces in the first stage are applied in such a way as to restrain the member against all deformation. The effect of this restraint is to set up a stress in the concrete equal, at any point, to the product of the unrestrained deformation and the modulus of elasticity.

In the second stage equal and opposite balancing forces are introduced as before. The stresses produced by these forces acting on the members are added to those set up in the first stage. The deformation of the member is that caused by the balancing forces.

Differential shrinkage in a composite member

EXAMPLE

A simple example of the use of the complementary method is shown when calculating the stresses in a composite member resulting from differential shrinkage of the two parts.

The composite section in Fig. 3.8 is made up of one of the Concrete Society standard precast bridge sections with an added top slab 1425 mm (54 in) wide and 185 mm (7 in) thick. It is required to calculate the stresses in the concrete if the limiting shrinkage of the top slab is 150 x 10⁻⁶ greater than the shrinkage of the precast section after casting the slab. It may be assumed that the modulus of elasticity of the concrete in both parts is 33 000 N/mm² (4.79 x 10⁶ lbf/in²).

The area of the precast section is 384 000 mm² (595 in²), and the second moment of area 69 200 x 10⁶ mm⁴ (166 658 in⁴). When the top slab has been added the area is increased to 628 000 mm² (973 in²) and the second moment of area to 185 500 x 10⁶ mm⁴.

If the top slab is restrained against contraction relative to the precast member due to differential shrinkage, it will develop a uniform tensile stress of 150 x 10⁻⁶ x 33 000 = 4.95 N/mm² (720 lbf/in²).

The imaginary force required to maintain this restraint is 4.95 x 1370 x 178 = 1 210 000 N (272 000 lbf) acting at the mid-depth of the slab.

The balancing force will therefore be of equal value, acting on the composite section so as to cause compression, and its moment about the centroid will be 1 210 000 x 526 x 10⁻⁶ = 637 kN m.

The stresses resulting from the balancing force may thus be calculated, and in the top slab they are added to the existing tensile stress of 4.95 N/mm². The final stresses are given in Fig. 3.8.

It will be seen that the general effect of differential shrinkage in a composite member of T section is to cause a small tensile stress in the top slab and at the bottom of the precast part of the member, while the main part of the latter is in

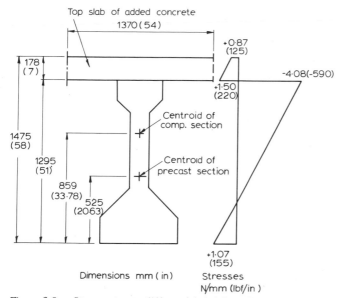

Figure 3.8   Stresses due to differential shrinkage in a composite section

compression. The British Code of Practice draws attention to the possible tensile stress due to differential shrinkage which may under certain conditions be significant and result in a tendency to earlier flexural cracking.

Advantages of complementary method

The first method described is more suitable for the normal type of reinforced or prestressed concrete member, but the complementary method has some advantages. It enables one to take account of a nonlinear distribution of strain due to shrinkage or creep, since under the initial imaginary condition of complete restraint the distribution of stress need not be linear, although the magnitude and position of the restraining and balancing forces will have to be found by integration.

Zienkewicz and Cruz [3.8] have developed the method for statically indeterminate structures. The balancing forces are, by differentiation, converted into a system of distributed axial and transverse loads, the effect of which can be analysed by the usual methods.

Simplified approximate calculation of loss of prestress

The prestress in a member is commonly expressed in terms of the stress in the tensioned steel and the effect of shrinkage and creep is assessed in terms of a loss of some of the tensile stress in this steel, as mentioned in Chapter 1. This method implicitly assumes that the tensioned steel is concentrated at the level of the centroid of its area and therefore introduces some inaccuracy, since the foregoing examples have shown that the loss of stress in the steel varies according to its position. Moreover, it does not take proper account of the effect of untensioned steel and may therefore seriously underestimate the loss of prestress in composite members with a large proportion of steel [3.9] unless the method is suitably modified [3.10]. Used with caution, however, it will be found sufficiently accurate for many routine design calculations.

If a shrinkage ϵ_{cs} causes a change of stress Δf_p in the prestressed steel, the corresponding change of stress Δf_c in the concrete at the level of the steel will be:

$$\Delta f_c = \Delta f_p \frac{A_p}{A_c} \left(1 + \frac{e^2}{i^2}\right).$$

This causes an elastic strain in addition to the shrinkage ϵ_{cs} and the total strain of the concrete must be equal to the strain of the steel.

Therefore

$$\frac{\Delta f_{\rm p}}{E_{\rm s}} = -\epsilon_{\rm cs} - \frac{\Delta f_{\rm p} A_{\rm p}}{E_{\rm c} A_{\rm c}} \left(1 + \frac{e^2}{i^2}\right),$$

$$\Delta f_{\rm p} = \frac{-E_{\rm s}\epsilon_{\rm cs}}{1 + \dfrac{\alpha A_{\rm p}}{A_{\rm c}} \left(1 + \dfrac{e^2}{i^2}\right)}$$

Since the denominator of the above expression is only a little greater than unity one can usually employ the simpler expression

$$\Delta f_{\rm p} = -E_{\rm s}\,\epsilon_{\rm cs}.$$

Similarly, it can be shown that the change of stress in the tensioned steel due to a specific creep $k_{\rm c}$ is given by

$$\Delta f_{\rm p} = \frac{-E_{\rm s} f_{\rm c} k_{\rm c}}{1 + \dfrac{\alpha A_{\rm p}}{A_{\rm c}} \left(1 + \dfrac{e^2}{i^2}\right)} \simeq -E_{\rm s} f_{\rm c} k_{\rm c},$$

where $f_{\rm c}$ is the initial stress in the concrete at the level of the steel.

If it is desired to calculate the change of stress in the concrete using the equivalent section which includes the tensioned steel, the above values must be multiplied by the factor β defined in Chapter 1 (p. 1.22).

EXAMPLE

The change of stress in the steel due to shrinkage in the section used in the previous examples (Fig. 3.5) is approximately $-193\,000 \times 300 \times 10^{-6} = -57.9\,{\rm N/mm^2}$ ($-8400\,{\rm lbf/in^2}$). The corresponding change of stress in the concrete is found to be $+0.87\,{\rm N/mm^2}$ ($+125\,{\rm lbf/in}$) at the bottom and $-0.20\,{\rm N/mm^2}$ ($-30\,{\rm lbf/in^2}$) at the top. These values may be compared with the result of the more accurate calculation in Fig. 3.5 from which the loss of compressive stress in the concrete at the bottom is seen to be underestimated by about 20%.

To calculate the loss of prestress due to creep by the approximate method, the initial stress in the concrete at the level of the centroid of the tensioned steel is found to be $9.17\,{\rm N/mm^2}$ ($1330\,{\rm lbf/in^2}$). The change of stress in the steel is therefore

$$-193\,000 \times 36 \times 10^{-6} \times 9.17 - -63.8\,{\rm N/mm^2}\ (-9300\,{\rm lbf/in^2}).$$

The corresponding change of stress in the concrete is $+0.96\,{\rm N/mm^2}$ ($+140\,{\rm lbf/in^2}$) at the bottom and $-0.22\,{\rm N/mm^2}$ ($-30\,{\rm lbf/in^2}$) at the top. The more accurate value for the loss of compressive stress, given in Fig. 3.7, is underestimated by about 10%.

Creep and relaxation of stress in steel

The time-dependent deformation of structural concrete members is almost entirely due to shrinkage and creep of the concrete. However, although most reinforcing steels are linear and elastic in their response to load, many of the steels used for prestressing undergo creep at normal temperatures when stressed to the high level necessary in practice. If a stressed length of the steel is maintained at constant length, the effect of creep is to reduce the stress, which eventually reaches a minimum value.

The relaxation for a particular steel may be determined experimentally by a test at constant length, and the result will be conservative for use in a prestressed member in which the steel will actually contract as a result of shrinkage and creep of the concrete. In the absence of experimental evidence the allowance for loss of stress in high strength steel suggested in the British Code is 10% when the stress is 80% of the tensile strength, decreasing to 8% when the stress is 70% of the strength, and then decreasing linearly to zero for a stress of 50% of the strength.

Experience indicates that the loss of stress in hard-drawn wire given a heat treatment under tension (termed stabilizing) is very much less and may be almost insignificant.

Cracked members

The imaginary load method for uncracked members may also be employed for cracked members by considering that the imaginary forces and balancing forces act on the equivalent cracked section discussed in Chapter 2 and illustrated in Fig. 2.3. It is not strictly accurate, however, since the position of the neutral axis of bending stress will be modified and the principle of superposition no longer applies. An alternative treatment is given, based on the conditions of equilibrium and strain.

In Fig. 3.9(a) ϵ_c and ϵ_s are the actual strain values at the top and at the level of the tensile reinforcement. The stress in the steel will be $E_s\epsilon_s$, but if the concrete has undergone a shrinkage ϵ_{cs}, the maximum stress in the concrete will be given by

$$f_c = E_c(\epsilon_c - \epsilon_{cs}).$$

If ξd is the depth of the neutral axis of stress in the concrete, it is seen from the diagram that

$$\frac{\epsilon_s + \epsilon_{cs}}{\epsilon_c - \epsilon_{cs}} = \frac{1 - \xi}{\xi}.$$

Hence

$$\frac{f_s + \epsilon_{cs} E_s}{f_c} = \frac{\alpha(1 - \xi)}{\xi}.$$

This replaces the strain relationship 2.1.2 derived for bending without shrinkage on page 38. The equilibrium equations, 1(a) and 1(b) are unaltered, and the resulting equation for ξ is found to be

$$\xi^3 - 3 \left[1 + \frac{M}{bd^2 \rho \epsilon_{cs} E_s} \right] \xi^2 - 6\alpha\rho \frac{M}{bd^2 \epsilon_{cs} E_s} \xi + 6\alpha\rho \cdot \frac{M}{bd^2 \rho \epsilon_{cs} E_s} = 0.$$

The stresses f_s and f_c may then be obtained from equation 1(b).

Figure 3.9(b) is the corresponding diagram for creep. If the initial stress in the concrete at the top is $f_{c\,sup}$ and k_c is the specific creep, the final stress in the concrete at the top will be

$$f_c = E_c (\epsilon_c - k_c f_{c\,sup}).$$

As a slight simplifying modification the creep is assumed to be zero at the level of the final neutral axis of strain and to increase linearly from this level to the top. The modified strain relationship now becomes

$$\frac{f_s}{f_c + k_c f_{c\,sup} E_c} = \frac{\alpha(1 - \xi)}{\xi}$$

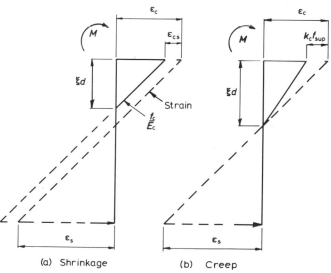

(a) Shrinkage   (b) Creep

Figure 3.9   Strain due to shrinkage and creep in a cracked member

and the equation for ξ is

$$\xi^3 - 3\left[4 + \frac{3M}{bd^2 \rho k_c f_{c\ sup} E_s}\right]\xi^2 + \left[3 - 6\alpha\rho \frac{M}{bd^2 \rho k_c f_{c\ sup} E_s}\right]\xi$$

$$+ 6\alpha\rho \frac{M}{bd^2 \rho k_c f_{c\ sup} E_s} = 0.$$

Deflexion of cracked members

The deflexion of a cracked member due to shrinkage or creep may be calculated from the curvature, derived from the stresses as described for uncracked sections. A simpler approach is usually employed in practice, however, whereby the additional time-dependent deflexion is obtained on multiplying the short-term·deflexion caused by the sustained load by a prescribed factor. The general value of this factor is 2 in both the American and European Concrete Committee Codes, but in the former it is reduced to take account of the stiffening effect of compressive reinforcement by multiplying by the factor $\{2 - 1.2\,(A_s'/A_s)\}$.

An alternative method is to use an effective (long term) modulus of elasticity E_{ce} to calculate the total deflexion. This is given by $E_{ce} = E_c/(1 + \phi)$ where E_c is the initial modulus and ϕ the creep coefficient. In calculating I the appropriate value of the modulus ratio (i.e. E_s/E_{ce}) should of course be used.

REFERENCES

[3.1]   Powers, T. C. (1968) Mechanisms of shrinkage and reversible creep of hardened cement paste. In *The Structure of Concrete* (Int. Conf. 1965). London: Cement and Concrete Assocn.

[3.2]   Carlson, R. W. (1938) Drying shrinkage of concrete as affected by many factors. *Proc. Amer. Soc. testing Materials*, **38, II**, 419-437.

[3.3]   Pickett, G. (1956) Effect of aggregate on shrinkage of concrete and hypothesis concerning shrinkage. *Proc. Amer. Concr. Inst.*, **52**, 581-90.

[3.4]   Troxell, G. E., Raphael, J. M., and Davis, R. E. (1958) Long-time creep and shrinkage tests of plain and reinforced concrete. *Proc. Amer. Soc. testing Materials*, **58**, 1101-1120.

[3.5]   Haller, P. (1940) *Shrinkage and creep of mortar and concrete*. Zurich: E.M.P.A.

[3.6]   Neville, A. M. (1963) *Properties of concrete*. pp. 312-318. London: Pitman.

[3.7]   Evans, R. H. and Bennett, E. W. (1962) *Prestressed concrete theory and design*. pp. 42-52. London: Chapman and Hall.

4
Non-linear deformation and failure

Redistribution of stress between concrete and steel

The non-linear stress-strain characteristic of concrete has been briefly mentioned in Chapter 1. It has important implications when considering the redistribution of stress between concrete and steel which occurs in a reinforced or prestressed concrete member at loads above the normal working level, and which must be taken into account in calculating the ultimate load.

The strain, as may be seen from the example in Fig. 4.1, increases more rapidly as the stress becomes greater and may have a value of above 2000×10^{-6} when the maximum stress is attained. Weaker concretes tend to develop greater strains. The strain will continue to increase, and the stress will decrease without actual collapse, although this part of the curve can only be obtained experimentally in a testing machine sufficiently stiff to accommodate the decreasing resistance of the specimen without causing sudden failure. The falling branch of the stress-strain curve is, however, of considerable significance where a structural member is stiffened by reinforcement or by adjacent concrete at a lower stress.

If one considers the stress in the steel reinforcement, which is bonded to the

[3.8]   Zienkewicz, O. C. and Cruz, C. R. (Sept. 1962) The 'equivalent load' method for elastic thermal stress problems with particular application to arch dams. *Proc. Inst. Civ. Engnrs.*, 23, 15-34.

[3.9]   Bennett, E. W. (Oct. 1963) Calculation of changes of stress in prestressed concrete. members. *Concr. Construct. Engng.* 58, 403-408.

[3.10]   Abeles, P. W. (June 1961) The effect of non-tensioned steel in prestressed concrete. *Reinf. Concr. Rev.* 5, 636-642. Amendment page 729 September 1961.

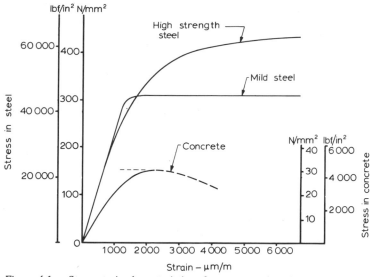

Figure 4.1　Stress-strain characteristics of concrete and steel

concrete and undergoes the same compressive strain, it will be seen from Fig. 4.1 that mild steel will have reached its yield point before the concrete reaches its maximum stress. In high strength steel, however, it is possible for the stress to continue to increase beyond the point of maximum stress in the concrete. Idealised forms of the stress-strain characteristics for steel and concrete, given in the British Unified Code, are shown in Fig. 4.2.

Failure of short reinforced columns in axial compression

Although it is customary to determine the compressive strength of concrete by the cube test in Great Britain and on the continent of Europe, this test does not give a result corresponding to the actual strength of concrete in a compressive member, because of the restraining effect of the platens of the testing machine. To avoid this problem the test specimen should have a height to breadth ratio not less than about 2. The standard American test on a 12 in. by 6 in. diameter cylinder is in this respect more satisfactory, but suffers from the disadvantage that the specimens have to be capped before testing.

The ratio of the cylinder strength or prism strength of concrete to the cube strength is not constant but may vary between 0.6 and 0.9 according to the properties of the concrete and the method of capping; moreover the strength obtained from tests on smaller specimens is slightly higher. However, a ratio of 0.8 can be considered satisfactory for general use. In the British Code the compressive strength of concrete in a column is assumed to be two-thirds of the

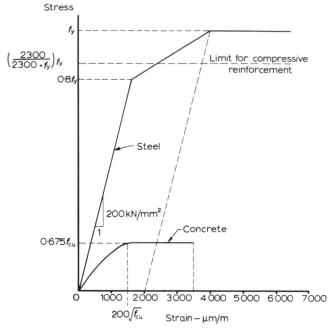

Figure 4.2   Idealised stress-strain characteristics of concrete and steel (British Code)

cube strength $(0.67 f_{cu})$ while in the American code the strength is assumed to be 0.85 times the cylinder strength $(0.85 f_c')$.

If the compressive reinforcement is of mild steel, and the yield stress is therefore attained before the concrete reaches its maximum stress, the ultimate load N_u will be given by

$$N_u = 0.67 f_{cu} A_c + f_y A_s',$$

where A_c is the area of concrete and A_s' the area of steel in compression.

The yield stress, f_y, for high strength steel is empirically assumed to be equal to the 0.2% proof stress, i.e. the stress at which the permanent strain is 0.2%. The total strain corresponding to a 0.2% proof stress of 420 N/mm² (60 000 lbf/in²) is about 400 x 10⁻⁶. In compression, therefore, this stress will not be attained in the steel when the strain is such that the stress in the surrounding concrete is at its maximum. Because of this, and the risk of buckling of the bars, discussed below, the compressive stress (which is considered to be developed in conjunction with the compressive strength of concrete) is reduced. The reduction factor proposed for compressive reinforcement in the British Code is 0.86; thus the formula for the failure of a column in axial compression is

$$N_u = 0.67 f_{cu} A_c + 0.86 f_y A_s'$$

The design strength N_d is calculated by dividing the characteristic cube strength of the concrete (f_{cu}) by the factor 1.5 and the characteristic strength of the steel by 1.15.* Thus

$$N_d = \frac{0.67 f_{cu} A_c}{1.5} + \frac{0.86 f_y A_s'}{1.15}$$

$$= 0.45 f_{cu} A_c + 0.75 f_y A_s'.\dagger$$

The above partial factors have been adopted in the British Unified Code following the recommendations of the European Concrete Committee (C.E.B.). The American code uses a single multiplying factor ϕ applied to the whole expression; the recommended value of ϕ is 0.70 for compression members with ties and 0.75 with spiral reinforcement.

The strength represented by the above expression will not be realised if the reinforcing bars are allowed to buckle outwards, causing spalling of the concrete cover and premature failure of the column (Fig. 4.3). This is prevented by the provision of lateral reinforcement, consisting of either separate binders (ties) or a continuous closely-spaced spiral. Binders should be bent and securely wired at the position of each main bar in order to give restraint in two directions. Rules for dimensioning them are given in the various codes of practice; although most of these have been established on the basis of practical experience, they have also been broadly confirmed by analysis and experiment [4.1].

Closely spaced reinforcement, such as a helix, has a further beneficial effect on the strength of a column. Concrete undergoes a considerable lateral strain in the last stages before compression failure which is restrained by this reinforcement, inducing a triaxial state of stress, and under these conditions the axial stress at failure is increased. The expansion is, however, sufficient to spall away the concrete cover outside the helix before failure of the restrained core, so that the strength of the column has to be calculated on the area of the core.

The following empirical expression has been proposed for the strength of an axially loaded column with helical reinforcement:

$$N_u = \frac{2}{3} f_{cu} A_{cc} + f_y A_s' + 80\,000\, A_{sh} \ (\text{lbf/in}^2)$$

*The significance of the various types of safety factor and the characteristic strength is discussed in Chapter 6.

†In the formula for the design of short braced axially loaded columns the coefficients are reduced by a further 10 per cent to allow for eccentricity due to construction tolerances.

where A_{cc} = area of core, i.e. internal diameter of helix,

A_{sh} = equivalent area of helix, i.e. volume of helix per unit length of
   column.

When the cover of concrete is large in relation to the diameter of the column
a higher value of N_u is given by the previous formula, and this should be used to
calculate the strength. American practice is to use a formula of the same type for
both spirally reinforced columns and columns with separate lateral ties, reducing
the strength of the latter by 15%, but the latest revision of the British Code
makes no special provision for helical reinforcement.

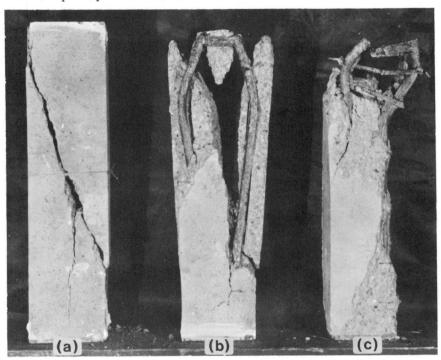

Figure 4.3   Failure of model columns (a) Unreinforced (b) Reinforced (c) Reinforced with
            lateral binders

Strength of reinforced concrete walls

Tests have shown that walls of plain concrete, short enough to preclude buckling
and subjected to a distributed axial load, can develop a compressive strength of
about $\frac{2}{3} f_{cu}$ similar to that of columns [4.2]. The strength was found to be
greater when the height/length ratio of the wall was less than 1.5 and was about
equal to the cube strength for a height/length ratio of 0.75. This phenomenon

was attributed to the restraining effect of the stiff beams through which the wall was loaded, an effect similar to that in a cube test, but acting principally in the direction of the length of the wall instead of in two dimensions and therefore affecting the strength slightly less.

The inclusion of a single layer of reinforcement did not sensibly increase the strength, but when reinforcement was placed in two layers near the wall faces and well tied together to form a cage, the increase in strength was more than could be attributed solely to the yield strength of the vertical steel. The additional increase appeared to be due to lateral restraint of the concrete as in spirally reinforced columns.

On the basis of this research the British Code permits an increase of the compressive stress in plain concrete walls varying linearly from zero when the height/length ratio is 1.5 to a maximum for a height/length ratio of 0.5. The amount of the increase is between 14% and 25% depending on the grade of the concrete. The strength of short braced axially loaded reinforced walls is calculated in the same way as that of columns.

Flexural failure

Progressive changes take place in the distribution of stress in both reinforced and prestressed concrete members between the unloaded state and the ultimate state where the load causes flexural failure. These are illustrated typically in Fig. 4.4. It will be recalled that a reinforced concrete member develops a tensile stress equal to the tensile strength of the concrete at a low load, and is therefore normally cracked at its working load. The prestressed member, on the other hand is normally uncracked at working load, although limited tensile stresses may be present in the concrete, and it is not until the member is overloaded that

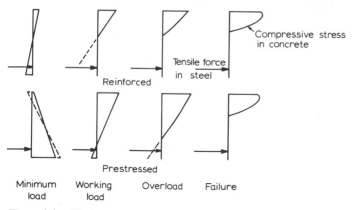

Figure 4.4   Typical stress distribution in beams loaded to failure

cracks will appear. If the load is further increased the cracks will widen and propagate towards the compression face of the member, whereas at this stage the cracks in the reinforced concrete member will grow wider but will not lengthen appreciably. In the last stages before failure of both types of member the position of the neutral axis may change since the stress in either the steel, the concrete or both, is no longer proportional to the strain. The life of the member is finally terminated by crushing of the concrete, or less commonly by fracture of the steel.

The non-linear stress-strain characteristic of the concrete results in the form of stress distribution shown in Fig. 4.4, sometimes rather inaccurately termed plastic. Since the strain in this region may be assumed to increase linearly from the neutral axis to the compression face, the stress distribution corresponds to part of the stress-strain characteristic of the concrete (illustrated in Fig. 4.1) turned through a right angle.

The position of the neutral axis may change just before failure in order to maintain equilibrium between the forces exerted by the steel and concrete. If at this stage the rate of increase of tensile strain of the steel is predominant, the neutral axis will move towards the compression face, whereas if the increase of strain of the concrete in compression has the greater influence, the neutral axis will move towards the tensile face. In the former case the section is termed under-reinforced, in the latter, which occurs with a greater area of steel, it is over-reinforced. Between the two the conditions are said to be balanced.

An under-reinforced member approaching its ultimate load is characterised by a relatively large deflexion and wide cracks (Fig. 4.5). This behaviour is desirable in that it affords a warning of overload; also where impact or shock loading has to be considered, due for example to earthquake or explosion, there is a high absorption of energy. Actual fracture of the steel, which occurs with very low percentages, is to be avoided, however, and a particularly dangerous situation arises if the flexural strength of the cracked section is less than the cracking load, so that sudden failure occurs at the instant of cracking. As a safeguard against this a minimum ratio of tensile reinforcement is usually specified, as for example $200/f_y$ (f_y = yield stress in lbf/in^2) for reinforced concrete in the American Code. In the British Code the ratio is given by the effective depth multiplied by 0.25% of the breadth for mild steel or 0.15% for high yield steel. In the latter the minimum area of steel required for prestressed concrete beams is 0.15% of the area obtained by multiplying the width of the bottom of the beam by its overall depth. A mimimum area of secondary reinforcement of 0.15% of the gross cross section for mild steel and 0.12% for high yield steel is specified for slabs; this is also necessary to control shrinkage cracking.

Over-reinforced members fail by sudden crushing of the concrete, with a

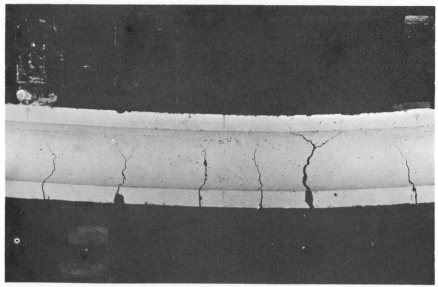

Figure 4.5   Failure of under-reinforced prestressed beam

small deflexion and narrow cracks. The suddenness and lack of warning of failure, particularly in prestressed concrete, are undesirable and the amount of reinforcement used in practice should preferably not exceed that required for a balanced section. The failure of an over-reinforced beam is illustrated in Fig. 4.6.

The flexural compressive zone

The analysis of a member failing in flexure will be based on the equilibrium and strain conditions as was done in considering uncracked and cracked sections with linear deformation in Chapters 1 and 2. The notation for the relevant dimensions and stresses for a rectangular section is given in Fig. 4.7. It is usual to assume that the shape of the stress distribution diagram can be represented by two parameters, namely the average stress and the position of the centre of compression. The former is written $\alpha_1\ \alpha_3\ f_{cu}$* where $\alpha_3\ f_{cu}$ is the maximum stress and f_{cu} the strength of the concrete; the latter is determined by the cube test in Britain, and if the cylinder test is used, as in the U.S.A. the values of the constants α_3 and $\alpha_1\alpha_3$ are correspondingly greater. The constant α_2 represents the distance of the centre of compression from the compressive face, expressed as a ratio of the depth of the compressive zone.

* The ratios k_1, k_2 and k_3 in the original literature have been amended to α_1, α_2 and α_3 in accordance with the International Standard Notation.

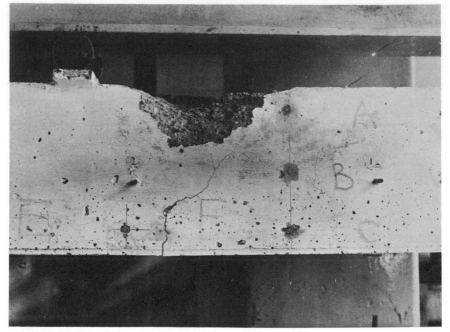

Figure 4.6   Failure of over-reinforced prestressed beam

A considerable amount of research has been carried out to establish the appropriate values of $\alpha_1 \alpha_3$ and α_2, notably by Hognestad et al. in America [4.3, 4.4], and by Rüsch in Germany. [4.5, 4.6] Actually these parameters are not constant but tend to be less for concrete of greater strength in which the non-linearity of the stress-strain characteristic is less pronounced. The American tests [4.3] were made on columns loaded with a controlled eccentric force so that the neutral axis always lay on the face, and they indicated values of $\alpha_1 \alpha_3$ (based on the cylinder strength) between 0.6 and 1.0 with α_2 between 0.35 and 0.5, later confirmed by the German tests. A method of deriving the flexural

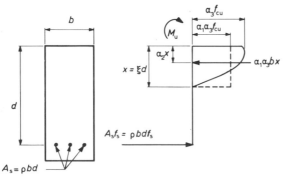

Figure 4.7   Notation for analysis of flexural strength

stress-strain curve from strain measurements on beams, developed by Bennett and O'Keeffe [4.7, 4.8] showed $\alpha_1 \alpha_3$ to be about 0.48 and α_2 0.38 based on the cube strength, when this was around 70 N/mm² (10 000 lbf/in²).

It must be appreciated that the maximum strain, and hence the stress distribution, will also depend on the shape of the concrete section in the compressive zone [4.6]. The compressive strain will continue to increase until the moment of resistance of the compressive force is a maximum, and it will be seen from Fig. 4.8 that, owing to the shape of the stress distribution, this will occur at a relatively low strain in a flanged section in which the centroid of the area of the compressive zone lies near the compressive face. On the other hand if the centroid is near the neutral axis, as in a member of square section bending about a diagonal, the moment of resistance will be greatest when the stress distribution is well developed, and includes an appreciable part of the descending branch of the stress-strain curve. It has been shown by Rüsch [4.6] that the rate of application and duration of loading have a further effect on the compressive zone; under sustained loading the maximum stress is reduced and the falling branch of the curve tends to become a plateau of constant stress.

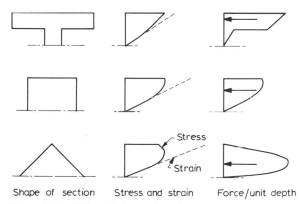

Shape of section   Stress and strain   Force/unit depth

Figure 4.8   Effect of shape of section on ultimate compressive stress and strain

Equilibrium equations for flexural strength

Referring to Fig. 4.7, the equilibrium equation for the forces on a rectangular section parallel to the axis is

$$f_s A_s = \alpha_1 \alpha_3 f_{cu} bx.$$

$$x = \frac{f_s A_s}{\alpha_1 \alpha_3 f_{cu} b},$$

$$\xi = \frac{\rho f_s}{\alpha_1 \alpha_3 f_{cu}} = \frac{(f_s/f_{su})}{\alpha_1 \alpha_3} \cdot \frac{\rho f_{su}}{f_{cu}}, \qquad \text{4.1.1(a)}$$

where f_{su} is the tensile strength of the steel.

From the condition of equilibrium of moments about the centre of compression or alternatively about the centre of the tensile reinforcement

$$M_u = f_s A_s (d - \alpha_2 x)$$

or

$$\alpha_1 \alpha_3 f_{cu} bx (d - \alpha_2 x).$$

Therefore

$$\frac{M_u}{f_{cu}bd^2} = \frac{\rho f_{su}}{f_{cu}} \left(\frac{f_s}{f_{su}}\right) (1 - \alpha_2 \xi) \qquad \text{4.1.1(b)}$$

or

$$\alpha_1 \alpha_2 \xi (1 - \alpha_2 \xi). \qquad \text{4.1.1(b)}'$$

Even when the compressive zone is defined by suitable values of the constants $\alpha_1 \alpha_3$ and α_2 there are still two unknown quantities in the above two equations, namely ξ the depth ratio of the compressive zone and f_s, the stress in the steel. A solution is therefore only possible either by making a further assumption concerning at least one of these two variables, or by considering the strain condition. The latter alternative will be examined first and then the former, which is generally adopted for practical purposes.

Strain relationship for flexure

The strain relationship is based on the usual assumption that the distribution of the average strain is linear, and it is further assumed that the corresponding change of strain in the tensile reinforcement or prestressed steel is equal to the average strain in the adjacent concrete multiplied by a bond factor, β. When the concrete is in compression and uncracked, this factor, β_1, may be taken as unity for bonded reinforcement, but when the concrete is in tension and assumed to be cracked it will have some other value, β_2.

Referring to the strain diagram in Fig. 4.9, ϵ_{cu} is the maximum compressive flexural strain in the concrete at failure, at which stage the total strain in the steel is ϵ_s. The increase in tensile strain of this steel corresponding to the tensile strain in the adjacent concrete is therefore

$$\epsilon_s - \epsilon_{se} - \beta_1 \epsilon_{ce}.$$

The average strain in the concrete at the level of the tensile steel is related to this by the factor β_2 and is

$$\frac{\epsilon_s - \epsilon_{se} - \beta_1 \epsilon_{ce}}{\beta_2}.$$

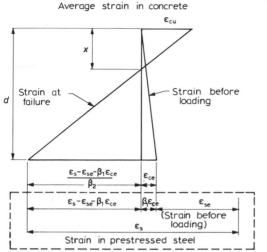

Figure 4.9 Strain diagram for analysis of flexural strength

The position of the neutral axis is now obtained from the strain distribution diagram

$$\xi = \frac{x}{d} = \frac{\beta_2 \epsilon_{cu}}{\epsilon_s - \epsilon_{se} - \beta_1 \epsilon_{ce} + \beta_2 \epsilon_{cu}}. \qquad 4.1.2$$

The accuracy of an analysis of flexural strength involving the above relationship obviously depends on the use of correct numerical values for the various terms. ϵ_{se} and ϵ_{ce} the effective strain in the steel and concrete due to the pre-stress, may be calculated with reasonable accuracy, while ϵ_s is related to the stress f_s by the stress-strain characteristic of the steel. One of the most significant terms in the expression, $\beta_2 \epsilon_{cu}$, is unfortunately extremely variable in both its factors. The ultimate flexural compressive strain of the concrete, ϵ_{cu}, is usually assumed to be constant with a value of about 3500×10^{-6} but experimental results have revealed a large range of values and it is also influenced by the shape of the section as has been shown above. The factor β_2 depends like β_1 on the bond between the steel and concrete, which is a property of the greatest importance.

In an uncracked member there will be no difference of strain between the bonded reinforcement and the adjacent concrete, and β_1 will be equal to 1. If, however, there are unbonded prestressing tendons, the strain must be constant over the full length, and β_1 will have a value less than unity, given theoretically by

$$\beta_1 = \frac{I_c}{M_c e_c} \int_0^l \frac{Me}{I} \, dx.$$

In this expression M the moment, I the second moment of area of the section, and e the eccentricity of the prestressing force are functions of x, the distance from one end of the member and have the values $M_c I_c$ and e_c at the critical section under consideration.

In a cracked member, as explained in Chapter 2, the local strain in well bonded steel at each crack will be greater than the average strain in the concrete, and the factor β_2 will therefore have a value greater than 1. The analysis of tests [4.7, 4.9] has revealed values of β_2 greater than 3 in some beams with pre-tensioned steel, although there is a large variation in these results. The value of β_2 for unbonded steel will again be less than 1 and a value equal to $0.8 \times \xi$ has been proposed [4.4]. It may be noted that the friction between the steel and the wall of the duct or the spacers, may produce an effect similar to a weak bond. Where the bond strength is moderate, as for example with post-tensioned tendons grouted in ducts, or reinforcing bars with a smooth surface or large diameter, β_2 will lie between these two extremes, and a value of 1 is often assumed.

In a beam with unbonded post-tensioned steel, the low ratio of the strain in the steel to the average strain in the concrete at the critical section results in a low rate of increase of the steel stress, and hence of the moment of resistance, in relation to the increase of deflexion and width of crack. This type of beam therefore fails at a lower load than a beam with well bonded steel and its

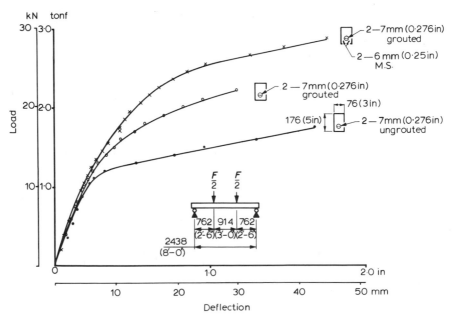

Figure 4.10   Load deflexion curves of prestressed beams with bonded and unbonded steel

behaviour after cracking is characterised by a larger deflexion and a small number of wide, typically bifurcated cracks, contrasting with the finer, well distributed cracks and lower rate of deflexion of a beam with bonded steel, as illustrated in Figs. 4.10, 4.11. For this reason, as well as the obvious advantage of protection against corrosion, the grouting of prestressing tendons is usually specified; however, non-bonded tendons are sometimes used to reduce costs where cracking is unlikely to occur, and it has been found that the inclusion of well bonded reinforcement in conjunction with non-bonded tendons largely eliminates the above undesirable features.

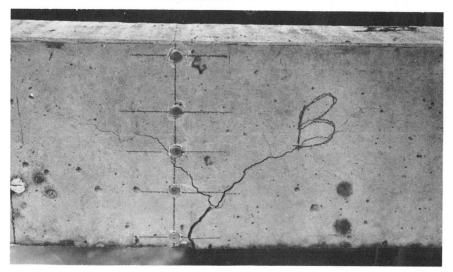

Figure 4.11   Cracks in prestressed beams with unbonded steel

Solutions of equations

The solution of the three equations 4.1.1(a), 4.1.1(b) and 4.1.2 is complicated by the fact that the stress f_s and the strain ϵ_s in the steel are related, the relationship being the stress-strain curve of the steel which will usually be non-linear at the values of stress and strain attained in a member at the point of flexural failure. The problem may be solved by iteration [4.4] or by graphical methods [4.10, 4.11] of which the one here described is due to Guyon.

The force equilibrium equation 4.1.1(a) is represented on the graph of f_s/f_u against ξ by a straight line passing through the origin and the point ($\rho f_{su}/f_{cu}$, $\alpha_1 \alpha_3$), as in Fig. 4.12. On the same diagram equation 4.1.2 is represented by plotting ξ for a series of values of f_s/f_{su}, obtaining the values of ϵ_s from the stress-strain curve of the steel. The solution of f_s/f_{su} and ξ is given by the

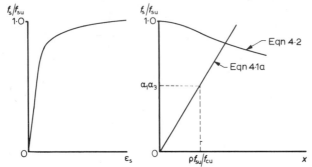

Figure 4.12   Graphical solution of flexural strength equation

intersection of the lines representing the two equations, and the ultimate moment M_u may then be obtained from equation 4.1.1(b).

Simplified methods

The above general method of analysis is laborious and requires the knowledge of data usually only available from an investigation under laboratory conditions. Consequently, although valuable in demonstrating the relative influence of the different variables, it is not suitable for use in practice, and a more direct approach is necessary.

The method adopted in most codes of practice is to dispense with the strain relationship by assuming that at failure either the tensile resistance of the steel or the compressive resistance of the concrete reaches an empirical maximum value. In the former event the section is under-reinforced; in the latter it is over-reinforced.

The British Code assumes that at flexural failure the average stress in the compressive zone is $0.6 f_{cu}$, ($\alpha_1 \alpha_3 = 0.6$) and that the centre of compression is located at the mid-depth of this zone ($\alpha_2 = 0.5$). If the maximum value of f_s, the stress in the tensile steel is f_{sb}, then equation 4.1.1(b) becomes

$$M_u = f_{sb} A_s \, (d - 0.5x)$$

In reinforced concrete members f_{sb} is considered to be equal to the yield stress or proof stress of the steel f_y so that from equations 4.1.1(a) and 4.1.1(b).

$$x = \frac{f_s A_s}{\alpha_1 \alpha_3 f_{cu} b} = \frac{f_y A_s}{0.6 f_{cu} b},$$

$$M_u = f_y A_s \left(d - 0.83 \frac{f_y A_s}{f_{cu} b} \right) .$$

The design flexural strength M_d is calculated by dividing the strength of the concrete and steel by the factors 1.5 and 1.15 respectively. Thus

$$M_{ud} = (0.87f_y)A_s z,$$

where the lever arm, z, is given by

$$z = d - 0.5\ \frac{(f_y/1.15)A_s}{0.6\,(f_{cu}/1.5)b}$$

$$= \left(1 - \frac{1.1f_y A_s}{f_{cu}\,bd}\right)d.$$

In under-reinforced prestressed members with low percentages of steel the stress developed may be assumed to be equal to the tensile strength, but it decreases as the percentage increases. The British Code recommendations imply the relationship between the depth of the neutral axis and the stress f_{pb} in the steel shown in Fig. 4.13, which corresponds to the graphical representation of the strain equation (Eqn. 4.1.2) in Fig. 4.12. f_{pb}/f_{pu} is also related to $\rho f_{pu}/f_{cu}$ by equation 4.1.1(a) and the ultimate moment is therefore given by the same expressions as for reinforced concrete, but with f_{pb} replacing f_y.

In a beam with unbonded tendons the term $\beta_2 \epsilon_{cu}$ in the strain equation is small and the term ϵ_{se} (the strain in the steel corresponding to the prestress) becomes more important than with bonded tendons. The depth of the neutral axis is therefore assumed to be related to the stress in the steel, expressed in the British Code as a proportion of the effective prestress, as in Fig. 4.13. The stress in unbonded steel is often little greater than the effective prestress, and the American Code specifies a value 15 000 lbf/in^2 (103 N/mm^2) greater than the

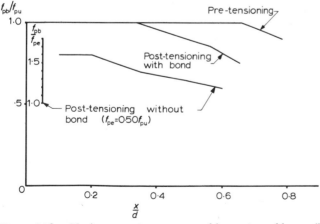

Figure 4.13   Maximum steel stress assumed in prestressed beams (British Code)

effective prestress. There is evidence that a more severe restriction is required on the minimum area of steel required in this type of member.

The maximum compressive resistance of the concrete is generally obtained from an empirically determined maximum depth of the compressive zone. The value of the ultimate moment is therefore constant for all values of the reinforcement ratio greater than that required to balance this maximum compressive resistance and may be obtained from equation 4.1.1(b).

The value adopted for the maximum depth of the compressive zone of a reinforced concrete section in the British Code is one half of the effective depth. Hence

$$M_u = 0.6 f_{cu} \frac{bd}{2} \left(d - 0.5 \frac{d}{2} \right)$$

$$= 0.225 f_{cu} bd^2.$$

For design

$$M_{ud} = 0.225 (f_{cu}/1.5) bd^2 = 0.15 f_{cu} bd^2.$$

According to the American Code the depth of the compressive zone for reinforced concrete under balanced conditions should be calculated from the strain relationship, assuming $\beta_2 \epsilon_{cu} = 3000 \times 10^{-6}$ and $E_s = 29 \times 10^6$ lbf/in². Since the terms ϵ_{se} and $\beta_1 \epsilon_{ce}$ are zero in equation 4.1.2

$$\xi = \frac{3000}{f_y/29 + 3000} = \frac{87\,000}{f_y + 87\,000} \text{ where } f_y \text{ is in lbf/in}^2,$$

$$= \frac{600}{f_y + 600} \text{ where } f_y \text{ is in N/mm}^2.$$

The compressive zone is represented in the American Code by a uniform stress of magnitude 0.85 times the cylinder strength f_c', extending to α times the depth of the neutral axis, where

$$\alpha = 0.85 \text{ for } f_c' \leqslant 4000 \text{ lbf/in}^2 \ (27.5 \text{ N/mm}^2),$$

$$\alpha = 1.05 - \frac{f_c'}{20\,000} \text{ for } f_c' \geqslant 4000 \text{ lbf/in}^2,$$

$$= 1.05 - \frac{f_c'}{138} \text{ where } f_c' \text{ is in N/mm}^2.$$

Flexural strengths calculated for rectangular sections by this method agree closely with those given by the British Code.

Because of the undesirable characteristics of an over-reinforced member, described above, it is required that the reinforcement ratio should not exceed 0.75 of the ratio corresponding to the balanced condition.

Comparison with test results

A characteristic curve for the flexural strength of a section of a member may be obtained from the relationship between $M_u/bd^2 f_{cu}$ and $\rho f_u/f_{cu}$ or $\rho f_{pu}/f_{cu}$ by combining the two equations 4.1.1(a) and 4.1.1(b). Hence

$$\frac{M_u}{f_{cu}bd^2} = \rho \frac{f_y}{f_{cu}} \left(\frac{f_s}{f_y}\right) \left\{1 - 0.83 \rho \frac{f_y}{f_{cu}} \left(\frac{f_s}{f_y}\right)\right\}.$$

This enables the results of tests to be compared with theory as in Fig. 4.14, in which data obtained at the University of Leeds [4.7, 4.9, 4.12] are shown in relation to the proposals of the British Code for reinforced and prestressed concrete. It is seen that the expressions for the ultimate moment are generally conservative, while the design strength, calculated with factored values of the strength of the steel and concrete is less than all the results. Some of the low results for beams with bonded post-tensioned steel were obtained in an investigation in which the quality of the bond was varied [4.7]; where this is in doubt it is recommended that the flexural strength should be calculated as for a beam with unbonded steel.

The flexural strength of members with unbonded steel, calculated according to the British Code, will almost always be underestimated by the requirement that the stress in the steel shall not exceed the prestress by more than 100 N/mm^2, as in Fig. 4.10(d). From the limited number of test results shown this appears to be a safe and more simple alternative to the use of Table 40 given in the Code.

Compressive reinforcement

The contribution of compressive reinforcement to the flexural strength depends on the stress attained, and therefore on the strain in the compressive zone of the member. Referring to Fig. 4.9, if the distance of the centre of the compressive reinforcement from the compressive face is d', the compressive strain in the reinforcement will be $\epsilon_{cu}(x - d')/x$. The stress may thus be calculated from the stress-strain characteristic of the steel.

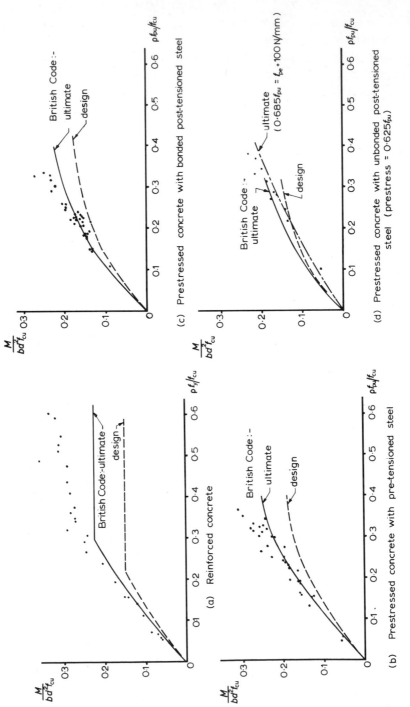

Figure 4.14 Results of tests of reinforced and prestressed concrete beams

The way in which the stress is related to the position of the steel in the compressive zone is illustrated in Fig. 4.15 which is based on the assumed maximum compressive strain and idealised stress-strain curves in the British Code. The stress in compressive reinforcement is limited as shown in Fig. 4.2 and it can be seen that the maximum value is likely to be attained when the reinforcement is located in the upper two fifths of the compressive zone, that is when $d' \leqslant d/5$ for a section controlled by failure of the concrete in compression, in which x is assumed to equal to $d'/2$.

Compressive reinforcement has little effect on the flexural strength of members in which failure is governed by the resistance of the tensile reinforcement, since it merely replaces part of the resistance of the concrete in compression, causing a negligible change in the lever arm. Its effect becomes marked when the ratio of tensile reinforcement is large, and the flexural strength can be increased by augmenting the controlling compressive resistance of the concrete. Using the maximum value $2300f_y/(2300 + f_y)$ for the stress in the compressive reinforcement as assumed in the British Code (Fig. 4.2), the strength formulae given above then become

$$M_u = 0.225\, f_{cu}\, bd^2 + \left(\frac{2300}{2300 + f_y}\right) f_y A_s' (d - d'),$$

$$M_{ud} = 0.15\, f_{cu}\, bd^2 + \left(\frac{2000}{2300 + f_y}\right) f_y A_s' (d - d')$$

$$\simeq 0.15\, f_{cu}\, bd^2 + 0.72\, f_y A_s' (d - d'),$$

where A_s' is the area of the compressive reinforcement.

The area of tensile reinforcement required for balanced conditions is increased and is given by

$$A_s = 0.3\frac{f_{cu}}{f_y}\, bd + \left(\frac{2300}{2300 + f_y}\right) A_s' \text{ for ultimate load,}$$

$$A_s = 0.23\frac{f_{cu}}{f_y}\, bd + \left(\frac{2000}{2300 + f_y}\right) A_s' \text{ for design.}$$

Prestressing tendons, for example pre-tensioned wires, in the compressive zone constitute a special case. The tensile prestress will be at least $0.55f_{pu}$ and it will be seen from Fig. 4.15 that the strain of the concrete in compression will reduce this stress but will probably not be sufficient to cause compressive stress.

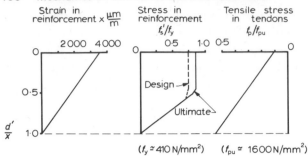

Figure 4.15   Stress in compressive reinforcement and tendons in the compressive zone

These tendons will therefore have an unfavourable effect on the ultimate moment, although where failure is controlled by the tensile reinforcement the reduction will usually be small enough to be neglected.

Compressive zone of non-rectangular section

The foregoing analysis is modified when the shape of the part of the section in flexural compression is not rectangular. This occurs for example in members having a compression flange or slab with the neutral axis lying outside it or in biaxial bending of a member of rectangular section.

Since, for the reason explained, the stress distribution under short-term loading varies with the shape of the compressive zone of the section, it appears that a complicated integration procedure might not necessarily be more accurate than a simpler approximation. On the other hand the stress distribution under sustained loading, which is of similar shape to that recommended by the European Concrete Committee and the British Code (Fig. 4.2) is more constant and the compressive moment of the concrete under sustained load could be obtained for a non-rectangular section by integration, using this stress distribution. However, comparative calculations and experimental data seem to indicate that the flexural strength can be calculated with reasonable accuracy by assuming an average compressive stress as for a rectangular section with the centre of compression located at the centroid of area of the compressive zone.

If the average compressive stress is $0.6\,f_{cu}$ the equilibrium equations for the flanged section shown in Fig. 4.16 are

$$f_s A_s = 0.6 f_{cu} b_w x + 0.6 f_{cu} (b - b_w) h_f$$

$$\text{or } x = \frac{f_s A_s}{0.6 f_{cu}} - \left(\frac{b - b_w}{b_w}\right) h_f$$

and $M_u = 0.6 f_{cu} b_w x(d - 0.5x) + 0.6 f_{cu} (b - b_w) h_f(d - 0.5\,h_f)$.

The stress f_s in the steel is equal to the yield stress f_y for reinforced concrete based on the tensile resistance, while for prestressed concrete it is either equal to the tensile strength or must satisfy a relationship with x such as that given in the British Code (Fig. 4.9).

When x reaches the assumed maximum value of $d/2$ and the ultimate moment is based on the strength of the concrete in compression

$$M_u = 0.225 f_{cu} b_w d^2 + 0.6 \left(1 - \frac{b_w}{b}\right)\left[\frac{h_f}{d} - 0.5\left(\frac{h_f}{d}\right)^2\right] f_{cu} bd^2.$$

The British Code simplifies the above formulae by assuming the compressive zone to be limited to the flange. The lever arm is then $d - h_f/2$ and the design flexural strength is given by

$$M_{ud} = (0.87 f_y) A_s \left(d - \frac{h_f}{2}\right)$$

with a maximum value of

$$M_{ud} = 0.4 f_{cu} b h_f \left(d - \frac{h_f}{2}\right)$$

The effect of the simplification can be slightly unconservative in the first of the above formulae, but conservative in the second, particularly where the breadth of the rib is substantial.

Supplementary reinforcement in prestressed concrete

The tendons in prestressed members have been seen to function as reinforcement and to provide the tensile resistance required in the cracked section. Their effect will be supplemented by that of any conventional reinforcement without prestress, and the flexural strength will thus be increased. A second important

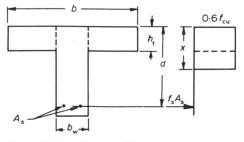

Figure 4.16   Flanged T section

function of supplementary reinforcement is the control of cracking, since it can be located close to the tensile face, whereas the tendons have sometimes to be placed nearer the centroid of the section.

Although there is a large difference between the prestress in the tendons and the slight compressive stress in the supplementary reinforcement before a member is loaded, it is usually possible for a high tensile stress to be attained in both at the ultimate load. This can be demonstrated by drawing the diagrams of strain and stress in the two types of steel, in the tensile zone as in Fig. 4.17, which has been based on the stress-strain curves in the British Code, assuming that the steel of the tendons has attained an ultimate elongation of 3%. Assuming the prestress to be $0.55 f_{pu}$ and f_{pu} to be about 1600 N/mm^2 (232 000 lb/in^2) the initial strain will be about 4400×10^{-6} and the approximate additional flexural strain required to cause failure will be $25\,600 \times 10^{-6}$. On the assumption that the flexural strain in the tensile zone increases linearly from zero at the neutral axis to this value at the level of the lowest tendon, the stress-strain curve of the steel, drawn in the diagram, represents the stress in a tendon in any other position, the stress being equal to the prestress at the level of the neutral axis. The stress corresponding to the assumed strain is also shown for supplementary reinforcement in which the stress is approximately zero at the neutral axis; two curves are given, one for high strength steel of the same quality as the tendons and the other for reinforcing steel of one quarter of the strength. The broken lines indicate the factored stresses for use in design.

It is seen from this diagram that at the ultimate moment, the stress developed in untensioned high strength steel can be considered equal to the tensile

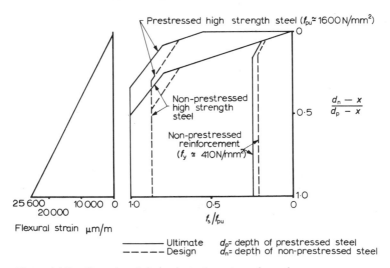

Figure 4.17   Stress in reinforcement of prestressed members

strength, provided it is located in the outer half of the tensile zone. More conservatively, the 0.2% proof stress may be assumed as in the earlier Code CP 115: 1959. Reinforcing steel may be considered to develop its yield stress when located in the outer three-quarters of the tensile zone.

A simple way of allowing for the effect of an area of reinforcement A_s which has a yield stress f_y, is to increase the area of the tendons by an amount $A_s f_y/f_{pu}$, adjusting the effective depth to take account of the position of this equivalent area of prestressed steel.

Combined compression and flexure

The equilibrium equations and strain relationship for the ultimate load conditions of a section subject to combined axial force and bending may be written, using the parameters $\alpha_1\alpha_3$ and α_2, as already defined, for the compressive zone of concrete. Referring to Fig. 4.18

$$N_u = \alpha_1\alpha_3 f_{cu} bx + f_{sc} A_s' - f_s A_s,$$

$$\frac{N_u}{f_{cu} bh} = \alpha_1\alpha_3 \left(\frac{x}{h}\right) + \rho' \frac{f_{sc}}{f_{cu}} - \rho \frac{f_s}{f_{cu}}. \qquad 4.2.1(a)$$

From the equilibrium of moments about the level of the tensile reinforcement

$$N_u \left[e + d - \frac{h}{2}\right] = \alpha_1\alpha_3 f_{cu} bx(d - \alpha_2 x) + f_{sc} A_s'(d - d'),$$

$$\frac{N_u e}{f_{cu} bh^2} + \frac{N_u}{f_{cu} bh}\left(\frac{d}{h} - \frac{1}{2}\right) = \alpha_1\alpha_3 \frac{x}{h}\left(\frac{d}{h} - \alpha_2 \frac{x}{h}\right) + \rho' \frac{f_{sc}}{f_{cu}}\left(\frac{d}{h} - \frac{d'}{d}\right). \qquad 4.2.1(b)$$

In writing the strain relationship it is assumed that the bond factor (β) for the steel is unity, hence

$$\frac{x}{d} = \frac{x}{h} \cdot \frac{h}{d} = \frac{\epsilon_{cu}}{\epsilon_{cu} + \epsilon_s} \qquad 4.2.2$$

As in simple flexure, failure may be controlled either by tension when the steel has attained its yield stress f_y or by compression of the concrete while the stress in the tensile reinforcement is at a level below the yield point. In the former case the depth of the compressive zone may be obtained by eliminating N_u from the equations 4.2.1(a) and 4.2.1(b) and solving the resulting quadratic for x.

The stress f_{sc} in the compressive reinforcement will usually be the maximum value, e.g. in the British Code $2300 f_y/(2300 + f_y)$, for the ultimate load, or

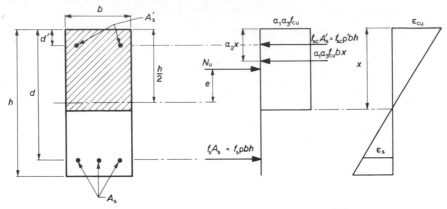

Figure 4.18   Internal forces and strain at failure by compression and flexure

$2000\,f_y/(2300 + f_y)$ for the design strength. This should be checked, however, when x has been computed, in the same way as for compressive reinforcement in flexure. N_u may then be obtained from equation 4.2.1(a).

The rather tedious arithmetic involved in the above calculations is avoided by the use of load-moment curves drawn for various values of the reinforcement ratios ρ and ρ', an example of which is given in Fig. 4.19. The section of the curve over which failure is controlled by tension lies nearest the horizontal axis and is drawn by calculating $N_u/f_{cu}bh$ and $N_u/f_{cu}bh^2$ from equations 4.2.1(a) and 4.2.1(b) for a series of values of x/h.

The tension-controlled section of the curve terminates at the 'balanced' point, at which the depth of the compressive zone is given by the strain relationship

$$\frac{x}{h} = \frac{\epsilon_{cu}}{\epsilon_{cu} + \epsilon_y}$$

In the British Code $\epsilon_{cu} = 0.0035$ and the strain assumed in the tensile steel at the commencement of yielding (Fig. 4.2) is

$$\epsilon_y = \frac{f_y}{200\,000} + 0.0020,$$

where f_y is the yield stress in N/mm^2.
Hence

$$\frac{x}{h} = \frac{700}{(d/h)(f_y + 1100)}.$$

The corresponding values of $N_u/f_{cu}bh$ and $N_u e/f_{cu}bh^2$ are readily obtained from equations 4.2.1(a) and 4.2.1(b).

When failure is controlled by compression of the concrete, calculation of N_u and $N_u e$ is more complicated, particularly where the tensile stress-strain curve of the steel becomes non-linear before the yield point. The load-moment curve for given reinforcement values may easily be drawn, however, by first using the strain relationship in conjunction with the stress-strain curve of the steel to obtain the value of the stress in the tensile reinforcement for a given value of x, and then substituting in equations 4.2.1(a) and 4.2.1(b) to obtain the co-ordinates of the corresponding point on the curve. Using the idealised stress-strain characteristic for steel from the British Code, the stress f_s in the tensile reinforcement is given in N/mm^2 by the expressions

$$f_s = \frac{700d/h}{x/h} - 700 \quad (f_s \leqslant 0.8f_y),$$

$$f_s = \frac{700d/h}{x/h} - 1100 \quad (0.8f_y < f_s < f_y).$$

The compression-controlled section of the curve in Fig. 4.19 has been derived by this method, and extends from the 'balanced' point to the point of intersection with the vertical axis, representing the axial compressive strength.

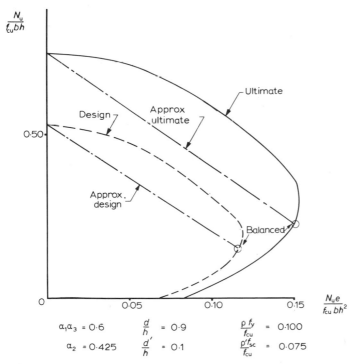

$$\begin{array}{lll} a_1 a_3 = 0.6 & \dfrac{d}{h} = 0.9 & \dfrac{p\,f_y}{f_{cu}} = 0.100 \\[2mm] a_2 = 0.425 & \dfrac{d'}{h} = 0.1 & \dfrac{p'f_{sc}}{f_{cu}} = 0.075 \end{array}$$

Figure 4.19   Load-moment curve for strength in compression and flexure

The value $\alpha_2 = 0.425$ has been used throughout to correspond more closely with the stress-strain curve for concrete in Fig. 4.2; this is also the value used in the former Code of Practice CP114:1957. The design strength curve, shown by a broken line, is obtained in the usual way by dividing the steel and concrete strength by the respective factors 1.15 and 1.5.

A conservative approximation for the strength of columns controlled by compression may be made without the use of curves by linear interpolation between the two points representing balanced failure and axial compression, as shown by the chain-dotted line in Fig.4.19. Then if N_{ou} is the axial ultimate load and N_{bu}, e_{bu}, the load and eccentricity for balanced failure

$$N_u = N_{ou} - \frac{N_u e}{N_{bu} e_{bu}} (N_{ou} - N_{bu}).$$

$$N_u = \frac{N_{ou}}{1 + \left(\dfrac{N_{ou}}{N_{bu}} - 1\right) \dfrac{e}{e_{bu}}}.$$

Compression and flexure of a member of circular cross-section

The circular cross-section is commonly used for columns and piles and its strength in compression and bending may be calculated by a method basically similar to that used for rectangular sections, although the analysis is slightly more complicated.

The section, stress distribution and strain distribution are shown with the relevant notation in Fig. 4.20. The reinforcing bars are equally spaced around

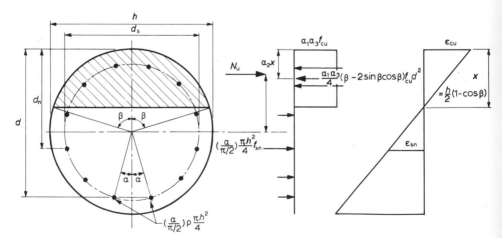

Figure 4.20   Compression and flexure of circular section

the circumference of a circle of diameter d_s so that two adjacent bars subtend an angle 2α at the centre. Thus the area of each pair of bars is $(2\alpha/\pi)\,\rho\pi h^2/4$ and the effective depth of the layer 'n' is given by

$$d_n = \frac{h}{2} + \frac{d_s}{2}\cos(2n-1)\alpha.$$

The compressive zone subtends an angle 2β at the centre; its depth x is therefore $h/2(1-\cos\beta)$ and its area $(\beta - \sin\beta\cos\beta)h^2/4$ where β is in radians.

The equilibrium equations are

$$\frac{N_u}{f_{cu}h^2} = \frac{\alpha_1\alpha_3}{4}(\beta - \sin\beta\cos\beta) + \left(\frac{2\alpha}{\pi}\right)\rho\frac{\pi}{4}\sum_{n=1}^{n=\pi/2\alpha} f_{sn}. \qquad 4.3.1(a)$$

$$\frac{N_u e}{f_{cu}h^3} = \frac{\alpha_1\alpha_3}{8}(\beta - \sin\beta\cos\beta)(1 - \alpha_2(1-\cos\beta))$$

$$+ \left(\frac{2\alpha}{\pi}\right)\rho\frac{\pi}{4}\sum_{n=1}^{n=\pi/2\alpha}\left(\frac{d_n}{h} - \frac{1}{2}\right) f_{sn}. \qquad 4.3.1(b)$$

The strain relationship for the bars at the level 'n' is

$$\epsilon_{sn} = \left\{\frac{2d_n}{h(1-\cos\beta)} - 1\right\}\epsilon_{cu},$$

$$\beta = \cos^{-1}\left\{1 - \frac{2d_n}{h}\left(\frac{\epsilon_{cu}}{\epsilon_{cu} + \epsilon_{sn}}\right)\right\} \qquad 4.3.2$$

The centre of compression will not necessarily lie at the centroid of the compressive zone owing to the effect of the shape of the section on the compressive stress distribution, and it is tentatively suggested that α_2 be given the value 0.5 for all values of β. In equations 4.3.1(a) and 4.3.1(b) the summation terms include both tensile and compressive reinforcement, the latter being indicated by a negative value of ϵ_{sn}.

For a given strength of steel and concrete and arrangement of reinforcement, the calculation of N_u and $N_u e$ may be done along the lines already described, but load-moment curves are to be preferred as the process is otherwise laborious. The values for the 'balanced' condition may, however, be readily obtained from the relationship

$$\beta = \cos^{-1}\left\{1 - \frac{2d}{h}\left(\frac{\epsilon_{cu}}{\epsilon_{cu} + \epsilon_y}\right)\right\}$$

where d is the effective depth of the lowest pair of bars.

This enables the interpolation method to be used for members in which failure is controlled by compression of the concrete.

Compression and bi-axial flexure

When a load acts eccentrically to both axes of a rectangular section, two moment equilibrium conditions have to be satisfied as well as the conditions of equilibrium of forces normal to the section. The moment may be calculated about the axes $x - x$ and $y - y$ (Fig. 4.21); thus if the area of concrete in compression is A_c', with the centre of compression at co-ordinates (e_{cx}, e_{cy}) and if A_s is the area of steel reinforcement at (e_{sx}, e_{sy}), the equations of equilibrium will be

$$N_u = \alpha_1\alpha_3 f_{cu} A_c' + \Sigma f_s A_s \qquad\qquad 4.4.1(a)$$

$$M_{xu} = N_u e_y = \alpha_1\alpha_3 f_{cu} A_c' e_{cy} + \Sigma f_s A_s e_{sy} \qquad\qquad 4.4.1(b)$$

$$M_{yu} = N_u e_x = \alpha_1\alpha_3 f_{cu} A_c' e_{cx} + \Sigma f_s A_s e_{sx}. \qquad\qquad 4.4.1(c)$$

In these equations the eccentricities (e) are positive in the upper righthand quadrant, and the stress in the steel (f_s), determined from the strain condition and stress-strain characteristics, is positive for compression.

An alternative approach is to make use of the condition that the force N, the resultant compressive resistance of the concrete and steel, and the resultant tensile resistance of the steel, must lie in one plane, i.e. the plane of bending. This replaces one of the two moment equations and the other is formulated from the equilibrium of moments in the plane of bending.

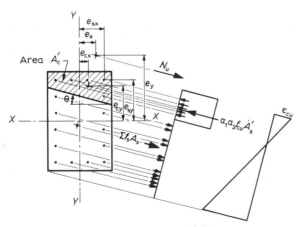

Figure 4.21   Compression and biaxial flexure

In either method the position and direction of the neutral axis has to be established by trial and error, and it must be appreciated that its direction will not necessarily be at right angles to the line joining the point of application of N with the centre of the section ($\theta = \tan^{-1}(e_x/e_y)$). Ramamurthy [4.13] found the error of this assumption to be between $+1°$ and $-6°$ for square columns with eight or more bars, but in rectangular sections the difference was greater, and for a load acting on a diagonal the neutral axis was approximately parallel to the other diagonal. The centre of compression of the concrete is usually assumed to be at the centroid of the compressive zone or of a smaller equivalent compressive zone bounded by a line parallel to the neutral axis as in the American code.

Many calculations have been made of the strength of biaxially loaded columns on varying the relevant parameters [4.4, 4.13, 4.22], and some of these have been verified by programmes of laboratory tests. Research has also aimed at establishing a relationship between the strength of these members and the better understood and more readily calculated strength in compression and uni-axial flexure [4.14 − 4.22]. This has generally been attempted by developing some form of the load-moment curve into a three-dimensional failure surface, sometimes represented two-dimensionally by a set of constant load contours.

The most commonly used failure surface is constructed for the orthogonal co-ordinates N_u, M_{xu}, M_{yu} as in Fig. 4.22. The lines of intersection of this surface with the vertical planes N_u, M_{xu} and N_u, M_{yu} are the load-moment curves for compression and uni-axial flexure about the axes $x − x$ and $y − y$ respectively. These are easily obtained and the ultimate moments M_{xu} and M_{yu} can be found for any biaxially eccentric value of N_u if the equation of the

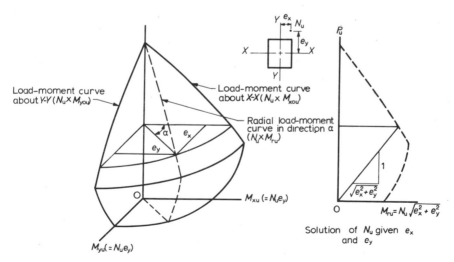

Figure 4.22 Failure surfaces for compression and biaxial flexure

appropriate constant load contour and the angle $\alpha(\alpha = \tan^{-1}(e_x/e_y))$ are known. A more common problem is that in which the bi-axial eccentricity (e_x, e_y) is fixed, and it is required to find N_u. This may be solved by using the two uni-axial load-moment curves and the constant load contours to construct the load-moment curve for bending in the radial plane, when the value of N_u may then be found by the simple construction illustrated in Fig. 4.22.

Several possible equations have been proposed to represent the constant load contours. For a member of square cross-section three points can be fixed on each contour since the load-moment curve is readily obtained for bending about a diagonal $(e_x = e_y)$, as well as about the principal axes. This has led Meek [4.20] to propose a bi-linear approximation for the contours, and Weber [4.23] has prepared curves based on the American code for use with this method, which is at present restricted to square columns with symmetrical reinforcement.

Constant load contours for rectangular sections can most easily be approximated by transferring one or both of the variables M_{xu} and M_{yu} so that the intercepts of each contour on the orthogonal axes are equal and the contour may be considered to be symmetrical about a line at 45° to these axes. Bresler [4.16] represents the above variables non-dimensionally as ratios of the uniaxial ultimate moments M_{xou} and M_{you}, and suggests a relationship of the form

$$\left(\frac{M_{xu}}{M_{xou}}\right)^n + \left(\frac{M_{yu}}{M_{you}}\right)^n = 1$$

The value of n is unfortunately not constant and has been shown to vary from about 1.1 to 4, sometimes in one column under varying conditions of eccentricity [4.19]. This method is recommended in the British Code.

Pannell [4.17] who first suggested the use of the $N_u \times M_{xu} \times M_{yu}$ surface as a tool for analysis and design, transforms the uni-axial load-moment curve $N_u \times M_{you}$ by multiplying by the ratio of the balanced moments about the principal axes M_{xob}/M_{yob} so that it becomes approximately the same as the curve $N_u \times M_{xou}$. The maximum radial deviation ζM_{xou}, of a constant load contour from a circular arc occurs at 45° to the two horizontal axes M_{xu}, $(M_{xob}/M_{yob})M_{yu}$, and the radial deviation of other points on the contour is given by the expression $\zeta M_{xou} \sin^2 2\theta$ where θ is the angle between the radius and one of the axes.

The shape of the forms of contours proposed by Meek, Bresler and Pannell are compared in Fig. 4.23 for typical positive and negative equal values of ζ. Meek's approximation is conservative while the curves of Pannell and Bresler do not differ greatly if the value of ζ is the same. However, Pannell's is the only

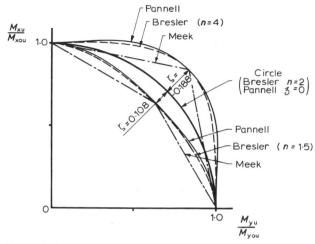

Figure 4.23   Approximations for constant load contours

method for rectangular sections which indicates the value of ζ; the latter has been obtained from a large number of computations and is presented graphically as a function of $N_u/f_{cu}bh$ and $\rho f_y/f_{cu}$. A further constant ψ has been computed to allow for the effect of different cover ratios. Hence

$$M_{xu} = \frac{(1 - \zeta \sin^2 2\theta) \cos \theta \, M_{xou}}{\psi} \, .$$

A different method of approximation proposed by Bresler [4.16] is based on the failure surface constructed for M_{xu}, M_{yu}, $1/N_u$. It is assumed that the failure surface is the plane between the required point $(M_{xu}, M_{yu}, 1/N_u)$, the corresponding points $(M_{xou}, 0, 1/N_{xu})$ and $(0, M_{you}, 1/N_{yu})$ on the vertical reference planes, and the point $(0, 0, 1/N_{0u})$ on the vertical axis. N_{xou} is the ultimate axial load when the moment M_{xu} is acting alone, N_{you} is the ultimate axial load when M_{yu} is acting alone, and N_{0u} is the ultimate axial load in the absence of any moment. This assumption yields the simple and convenient relationship

$$\frac{1}{N_u} = \frac{1}{N_{xou}} + \frac{1}{N_{you}} - \frac{1}{N_{0u}} \, .$$

Although only strictly true for conditions of linear deformation this has been found to give surprisingly good results for reinforced concrete.

REFERENCES

[4.1]　Bresler, B. and Gilbert, P. H. (Nov. 1961) Tie requirements for reinforced concrete columns. *Proc. Amer. Concr. Inst., 58,* 555-570.

[4.2]　Seddon, A. E. (1956) *The strength of concrete walls under axial and eccentric loads. Symposium on the strength of concrete structures.* London: Cement & Concr. Assoc.

[4.3]　Hognestad, E., Hanson, N. W. and McHenry, D. (Dec. 1955) Concrete stress distribution in ultimate strength design. *Proc. Amer. Concr. Inst.,* 27, 455-479.

[4.4]　Mattock, A. H., Kriz, L. B. and Hognestad, E. (Feb. 1961) Rectangular concrete stress distribution in ultimate strength design. *Proc. Amer. Concr. Inst., 57,* 875-928.

[4.5]　Rüsch, H. (1955) Versuche zur Festigkeit der Biegedruckzone (Investigation of the strength of the flexural compressive zone). *Deutscher Ausschuss für Stahlbeton,* 120, 5594.

[4.6]　Rüsch, H. (July 1960) Researches toward a general flexural theory for structural concrete. *Proc. Amer. Concr. Inst., 57,* 1-28.

[4.7]　Bennett, E. W. and O'Keefe, J. D. (Dec. 1958–Feb. 1959) The influence of bond on the flexural strength of prestressed concrete beams. *Civil Engng. (Lond.),* 53, 1391-1393; 54, 89–90, 209-212.

[4.8]　O'Keefe, J. D. and Bennett, E. W. (Dec. 1959) Nouvelle méthode pour déterminer la répartition des contraintes de compression dans les pièces fléchies en béton armé et en béton précontraint. (New method of determining the distribution of flexural compressive stresses in reinforced and prestressed concrete members). *Travaux,* 43 an. No 302, 680-683.

[4.9]　Bennett, E. W. and Ramesh, C. K. (Dec. 1959) An experimental study of some prestressed beams of varying tendon diameter. *J. Inst. Engnrs. (India)* 40, 241-250.

[4.10]　Guyon, Y. (London 1960) Prestressed Concrete. *Contractor's Record,* 2, 378-389.

[4.11]　Billet, D. F. and Appleton, J. H. (1954) Flexural strength of prestressed concrete beams. *Proc. Amer. Concr. Inst.,* 50, p. 837.

[4.12]　Evans, R. H. (Dec. 1943) The plastic theories for the ultimate strength of reinforced concrete beams. *J. Inst. Civil Engnrs.,* 21, 98-121.

[4.13]　Ramamurthy, L. M. (1966) Investigation of the ultimate strength of square and rectangular columns under biaxially eccentric loads. *Symp. on Reinforced Concr. Columns, Amer. Concr. Inst.* Paper No. 12,

[4.14]　Au, T. (Feb. 1958) Ultimate strength design of rectangular concrete members subject to unsymmetrical bending. *Proc. Amer. Concr. Inst.,* 54, 657-674.

[4.15]　Au, T. and Parbaccus, A. (Dec. 1958) Biaxially loaded reinforced concrete columns. *Proc. Amer. Soc. Civil Engnrs.,* 84, Struct. Div., Paper 1865.

[4.16]　Bresler, B. (Nov. 1960) Design criteria for reinforced columns under axial load and biaxial bending. *Proc. Amer. Concr. Inst.,* 57, 481-490.

[4.17]　Pannell, F. N. (July 1960) The design of biaxially loaded columns by ultimate load methods. *Mag. Concr. Res.,* 12, 99-108.

[4.18]　Furlong, R. W. (March 1961) Ultimate strength of square columns under biaxially eccentric load. *Proc. Amer. Concr. Inst., 57,* 1129-1140.

[4.19] Pannell, F. N. (Jan. 1963) Failure surfaces for members in compression and biaxial bending. *Proc. Amer. Concr. Inst.,* **60,** 129-140.

[4.20] Meek, J. L. (Aug. 1963) Ultimate strength of columns with biaxially eccentric loading. *Proc. Amer. Concr. Inst.,* **60,** 1053-1064.

[4.21] Aas-Jakobsen, A. (March 1964) Biaxial eccentricities in ultimate load design. *Proc. Amer. Concr. Inst.,* **61,** 317-333.

[4.22] Fleming, J. F. and Werner, S. D. (March 1954) Design of columns subject to biaxial bending. *Proc. Amer. Concr. Inst.,* **62,** 327-342.

[4.23] Weber, D. C. (Nov. 1966) Ultimate strength design chart for columns with rectangular bending. *Proc. Amer. Concr. Inst.,* **63,** 1205-1230.

5

Other modes of failure

Failure involving a length of member

The types of failure discussed in the previous chapter are localised at one critical section of a member, and provided the latter is statically determinate, its strength depends only upon the dimensions and properties of the materials at that section. The following chapter is devoted to some further modes of failure in which part or all of the length of the member is involved. This class includes failures due to shear and torsion, together with those caused by lateral instability of the member.

Cracking and shear failure of members without web reinforcement

It has been shown in Chapter 1 that, in both reinforced and prestressed concrete members, shearing forces are associated with principal tensile stresses inclined to the axis. When the shear exceeds a certain level, these stresses result in the formation of an inclined crack. This may happen in two ways [5.1]. Where there is an appreciable bending moment, a flexural crack may commence normal to the axis at the tensile face and later develop into an inclined crack, or an inclined crack may link up with an existing system of flexural cracks (Fig. 5.1). Cracks of this type are sometimes known as flexure-shear cracks. On the other hand, one or more inclined cracks may suddenly appear in the region of the mid-depth without any previous flexural cracking, particularly in prestressed members with thin webs. These may be termed web-shear cracks, which are also illustrated in Fig. 5.1. The occurrence of shear cracks in a member without shear reinforcement may be rapidly followed by failure as the inclined crack extends to the compression face so that the member is severed. This is known as the diagonal tension mode of shear failure (Fig. 5.2), and in calculations it is generally assumed that it occurs at the same load as inclined cracking.

Web-shear cracking

The shearing force at which web-shear cracking occurs is usually assumed to be governed by a limiting value of the principal tensile stress in the concrete. The calculation for a prestressed member uncracked by flexure may thus be done by the standard method described in Chapter 1 (p. 27ff). This is the basis of the recommendation in the British Code, where a simplified formula is given for the shearing force, V_{co}, corresponding to a given principal tensile stress.

$$V_{co} = 0.67\, b_w \sqrt{(f_t^2 + 0.8\, f_{cp} f_t)}.$$

b_w = breadth of web of member

h = overall depth of member

f_{co} = compressive prestress at centroid of section.

The expression has been based on the assumption of a maximum shear stress of $V_{co}/0.67 b_w h$ which is correct for a rectangular section but somewhat low for a flanged section. This, together with the reduced value of $0.8 f_{cp}$ for the

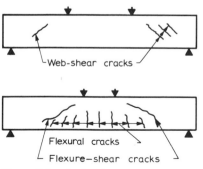

Figure 5.1   Inclined cracks due to shear

prestress, compensates for the fact that in flanged sections the maximum principal tensile stress does not always occur at the centroid.

The limiting value of the principal tensile stress is often taken to be the tensile strength of the concrete, as in the above formula in which $f_t = 0.24\sqrt{f_{cu}}$

Figure 5.2   Shear failure of a reinforced concrete beam by diagonal tension

where f_{cu} is the cube strength. Cracking tends to occur at a lower value, however, when the principal compressive stress is high, and this is taken into account in the European Concrete Committee's Recommendations, in which the expression for the limiting principal tensile stress is

$$\frac{f_{t\ max}}{f_{ct}} \leqslant 1 - 0.8\frac{f_{c\ max}}{f_c'}.$$

$f_{t\ max}$ = principal tensile stress,

$f_{c\ max}$ = principal compressive stress,

f_{ct} = tensile strength of concrete,

f_c' = compressive strength of concrete.

The American Code includes a simplified approximation of the formula for the web-shear cracking of a prestressed member [5.1], given by

$$V_{co} = b_w d(3.5\sqrt{f_c'} + 0.3 f_{cp}).$$

b_w = breadth of web of member,

d = effective depth of tendons,

f_c' = cylinder strength of concrete in lbf/in² (for f_c' in N/mm² the coefficient 3.5 is changed to 0.29),

f_{cp} = compressive prestress at centroid of section.

In both this and the previous formula the shearing force V_{co} is increased by $P \sin \theta$, the vertical component of the prestressing force.

Web-shear cracking is less common in reinforced concrete than in prestressed concrete, owing to the greater web thickness and earlier occurrence of flexural cracking. It has been customary to use the conventional expression $v = V/zb_w$ for uncracked as well as cracked members, since the shear stress v is numerically equal to the principal tensile stress at the level of the neutral axis. However, in view of the doubtful significance of the lever arm z and the essentially empirical character of shear formulae, it has become more usual to base calculations on the average shear stress over the effective section ($v = V/b_w d$) and to make a compensating adjustment to the value of the shear stress at which cracking is assumed to occur.

Flexure-shear cracking in reinforced concrete

The load at which a flexure-shear crack will appear depends on several factors, not included in the conventional shear stress formula ($v = V/zb$) which is therefore inadequate and sometimes unsafe. The percentage of main tensile reinforcement has an important effect, by controlling the rate of increase of width at the base of a flexural crack which leads to a flexure-shear crack and by transmitting some of the shear by dowel action. The relative magnitude of moment and shear, as indicated by the ratio M/Vd is also significant since the extent of flexural cracking will increase with this ratio. Finally, the strength of the concrete is of obvious importance and the significant parameter is considered by most research workers to be the tensile strength, or the compressive strength raised to a power between $\frac{1}{3}$ and $\frac{1}{2}$.

The American Code gives a semi-empirical expression for the shear cracking load of a reinforced concrete beam, which takes account of the above variables. The unfactored form of this expression is

$$v_c = 1.9\sqrt{f_c'} + \frac{2500\rho Vd}{M} \quad \text{(for stresses in lbf/in}^2)$$

and

$$v_c = 0.16\sqrt{f_c'} + \frac{17.2\rho Vd}{M} \quad \text{(for stresses in N/mm}^2),$$

where ρ is the tensile reinforcement ratio, $\dfrac{A_s}{bd}$

The value of v_c must not exceed $3.5\sqrt{f_c'}$ lbf/in² ($0.29\sqrt{f_c'}$ N/mm²) which corresponds to the web-shear cracking load as given by the prestressed concrete formula above.

The British Code has now introduced a new set of shear clauses based on the report of a study group of the Institution of Structural Engineers [5.2]. The specified factored values of the average shear stress v_c for a range of values of ρ and concrete strength are given in Table 5.1; these have been compiled from a comparison of several theories and checked against experimental results, and an example of the correlation is given in Fig. 5.3. A conservative approximation of the tabulated values may be obtained from the expression

$$0.7\,(100\,\rho)^{1/3} + \frac{f_{cu}}{100} - 0.4 \quad \text{(for stresses in N/mm}^2)$$

and

$$100\,(100\,\rho)^{1/3} + \frac{f_{cu}}{100} - 60 \quad \text{(for stresses in lbf/in}^2).$$

Table 5.1   *Ultimate average shear resistance of concrete* v_c *(N/mm²) (British Code)*

Tensile reinforcement ratio $\rho = \dfrac{A_s}{bd}$ per cent	Cube strength of concrete N/mm²			
	20	25	30	40 or more
0.25	0.35	0.35	0.35	0.35
0.50	0.45	0.50	0.55	0.55
1.00	0.60	0.65	0.70	0.75
2.00	0.80	0.85	0.90	0.95
3.00	0.85	0.90	0.95	1.00

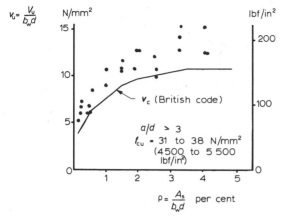

Figure 5.3   Shear strength of beams failing by diagonal tension [5.2]

When an inclined crack leads to diagonal tension failure of a member without web reinforcement, it is often accompanied by splitting of the concrete from the lower end of the crack along the tensile reinforcement to the support, a tendency which would be checked by stirrups. In view of the weakness at this point the shear strength cannot be increased indefinitely by increasing the tensile reinforcement, and it is in fact limited by the maximum value of $v_c = 3.5\sqrt{f_c'}$ lbf/in^2 ($0.29\sqrt{f_c'}$ N/mm^2) in the American Code or in the British Code by a maximum value of v_c corresponding to a tensile reinforcement ratio of 3%, and a concrete cube strength of 40 N/mm^2 (6000 lbf/in^2).

Effect of axial load on flexure-shear cracking

An axial load, by modifying the principal tensile stress in the concrete, will obviously have a considerable effect on the cracking shear, a tensile load causing earlier and a compressive load later cracking. Since the axial force appears to influence the flexure-cracking load mainly by its effect on the height of the flexure cracks, the American Code has introduced the concept of an equivalent moment M_e derived from the equilibrium conditions of a section subject to a moment M and an axial force (positive when compressive), where

$$M_e = M - \frac{N(4h - d)}{8}.$$

The moment M_e replaces M in the flexure-shear cracking formula quoted above, with the proviso that

$$v_c \leqslant 3.5\sqrt{\{f_c' (1 + 0.002 \, N/A)\}} \qquad \text{(stresses in lbf/in}^2\text{)}$$

where A is the gross area of the section.

This method has been found to give a satisfactory lower bound to the results of tests of members subject to axial compression, but to be unconservative for some members subject to axial tension. Because of this, together with the fact that the formula is not convenient to use, alternative formulae have been proposed [5.1]. For members subject to axial compression

$$v_c = \frac{2 f_c'}{1 - 0.0008 \, N/A} \qquad \text{(stresses in lbf/in}^2\text{)}.$$

For members subject to axial tension (*P* negative)

$$v_c = 2(1 + 0.002 \, N/A)\sqrt{f_c'} \qquad \text{(stresses in lbf/in}^2\text{)}.$$

Only the second of these formulae has been adopted in the 1971 American Code.

Mattock [5.5] has suggested an approach in which the American flexure-shear formula for a prestressed beam is developed so as to apply to reinforced concrete members with or without axial load. The tensile reinforcement ratio ρ was found to have an important influence.

Flexure-shear cracking in prestressed concrete

The formula for the flexure-shear cracking load in a prestressed member, derived empirically by ACI-ASCE Committee 326 for the American Code [5.3], is based on the experimental observation that flexure-shear cracking appears to occur at the flexural cracking load plus an additional shear which is a function of the strength of the concrete and the dimensions of the section.

The flexural cracking moment must first be calculated. Flexural cracking is assumed to occur when the tensile stress is $6\sqrt{f_c'}$ lbf/in², which is equivalent to $0.45\sqrt{f_{cu}}$ N/mm² if $f_c' = 0.8 \, f_{cu}$.

If M_r is the total flexural cracking moment (including self weight) the corresponding total shear is VM_r/M where M/V is the moment/shear ratio at the critical section. The additional shear was found to be conservatively represented by the expression $0.6 \, b_w d \, f_c'$lbf/in², so that the average shear stress for flexure-shear cracking is

$$v_{cr} = \frac{V_{cr}}{b_w d} = 0.6 \sqrt{f_c'} + \frac{V_g V_q M_r/M_q}{b_w d} \qquad \text{lbf/in}^2,$$

but need not be taken less than $1.7\sqrt{f_c'}$ where V_g is the shear due to dead load and V_q and M_q are the shear and moment due to live load. The corresponding factored expression for the design stress for flexure-shear cracking in the British Code is

$$v_{cr} = \frac{V_{cr}}{b_w d} = \left(1 - 0.55 \frac{f_{pe}}{f_{pu}}\right) v_c + \frac{M_o}{b_w d^2} \cdot \frac{Vd}{M}$$

but not less than $0.1\sqrt{f_{cu}}$,
where

$$M_o = \frac{0.8 f_{cp}}{y_{inf}}$$

I = moment necessary to produce zero stress in the concrete at the depth d,

f_{cp} = stress due to prestress only at the extreme tensile fibre,

y_{inf} = distance of extreme tensile fibre from centroid of section,

I = second moment of area of section,

f_{pe} = effective prestress in steel after all losses have occurred, not to be considered to be greater than $0.6 f_{su}$,

f_{pu} = tensile strength of prestressed steel,

v_c = ultimate average shear stress for reinforced concrete (Table 5.1).

Shear-compression failure

The occurrence of shear failure by diagonal tension as an immediate result of inclined cracking of either type may be prevented by shear reinforcement or by the local compressive stresses set up in the region of concentrated loads, particularly by the inclined compression when the load is close to the support. The shear strength of a member is found to increase markedly when the shear span a, that is, the distance between the support and the nearest concentrated load, becomes less than about three times the effective depth. In the absence of web reinforcement this phenomenon is much less pronounced when shear is applied at the side by a secondary member, than when it arises from a concentrated load bearing on the compressive face of the member [5.6].

The mode of failure which eventually follows flexure-shear cracking under these conditions is a modified form of flexural failure known as shear-compression (Fig. 5.4). Various theories have been proposed, most of them involving somewhat sophisticated models of the strain compatibility condition resulting in a modification of the stress parameters $\alpha_1 \alpha_3$ and α_2 for the compressive zone, as defined in the previous chapter (p. 93). A recent comparison of a number of these theories [5.2] has shown a good degree of agreement between the predicted shear-compression moments.

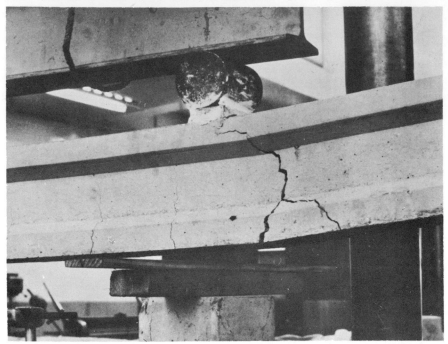

Figure 5.4   Shear compression failure of a prestressed beam

The American formula for the flexure-shear cracking of reinforced concrete makes a limited allowance for the influence of the ratio M/Vd. The British proposal [5.2] is that the shear resistance of the concrete, v_c, should be increased by multiplying by the ratio $2d/a$ when the shear span a is less than $2d$, provided v does not exceed the maximum value discussed below. This increase, which is limited to loads applied to the compression face and applies also to corbels and brackets, is compared with test results in Fig. 5.5.

Figure 5.6 shows the effect of the ratio M/Vd on the mode of shear failure. When the ratio is large the moment will reach the flexural strength before the shear strength is attained. For somewhat lower values of the ratio, flexure-shear cracking will lead to shear failure by diagonal tension before the flexural strength has been reached, and for still lower values shear cracking will not result in immediate failure, but there will be a higher ultimate shear corresponding to the shear-compression moment.

The increase of shear strength obtained with very short shear spans is limited by the occurrence of a further type of failure associated with distortion and crushing of the concrete. For calculation this is defined by an upper limit to the average shear stress, the factored value recommended in the British Code being $0.75\sqrt{f_{cu}}$.

Shear failure of members with web reinforcement

It has been observed from experiments that the action of web reinforcement, either stirrups or inclined bars, is to permit a redistribution of the internal forces after inclined cracking has occurred. Not only does this reinforcement contribute directly to the shear resistance in the region of the inclined cracks, but, by arresting the widening and propagation of these cracks towards the compression face, it also prevents shear failure by diagonal tension and enables the uncracked concrete near the compressive face to contribute to the shear resistance. Failure is often by shear-compression and attempts have been made to develop shear-compression theories to include the effect of web reinforcement [5.2].

The classical (Mörsch) formula derived in Chapter 2 (p. 53f) may be written

$$v = \frac{V}{b_{w}z} = \beta\rho_{v}f_{sv}$$

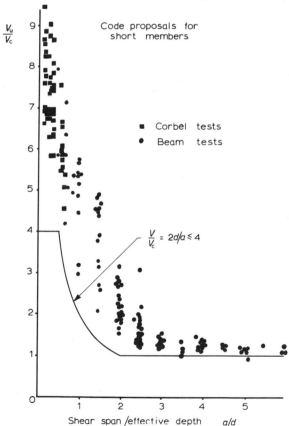

Figure 5.5   Effect of shear span ratio on shear strength [5.2]

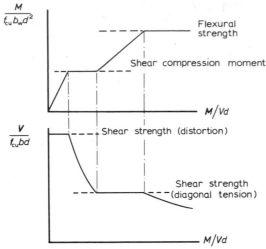

Figure 5.6   Effect of moment/shear ratio on mode of failure

or approximately

$$v = \frac{V}{b_{\mathrm{w}} d} = \beta \rho_{\mathrm{v}} f_{\mathrm{sv}}$$

It has been found to be over-conservative when compared with the results of tests, as it makes no allowance for the contribution of the uncracked concrete in the compressive zone to the total shear resistance, nor for the further increase in strength due to aggregate interlock at the cracks and dowel resistance of the reinforcement.

It was proposed in 1962 in America by ACI-ASCE Committee 326 [5.4] that the shear strength should be calculated by assuming the contribution of the web reinforcement to be given by the Mörsch formula with the steel stress f_{sv} equal to the yield stress f_{y}, and by assuming that the additional strength provided by the concrete is equal to the shear cracking load. Thus

$$v_{\mathrm{u}} = \frac{V_{\mathrm{u}}}{b_{\mathrm{w}} d} = \beta \rho_{\mathrm{v}} f_{\mathrm{y}} + v_{\mathrm{c}}.$$

The ACI expression was based on a study of the ultimate loads of beams tested to failure, but the effect of the shear cracking load can also be seen from the diagram of stress in the web reinforcement against the total shear (Fig. 5.7).

Results of this type have been presented by Leonhardt both for reinforced concrete [5.7] and for prestressed concrete [5.8]. Little stress can be detected until the bar or stirrup is intersected by an inclined crack, but beyond this point the increase of stress with respect to shear is at first about the same as that given by the classical formula. If this rate of increase were to continue up to failure, as in Fig. 5.7, the shear contribution of the concrete would be equal to the inclined cracking load, but this is not always so, and under some conditions it may be less.

The above type of formula, with the resistance of the concrete equal to v_c, has been adopted in the British Code for the design of shear reinforcement for reinforced and prestressed concrete members, and has been shown to give a conservative prediction of the strength.

The 1970 Recommendations of the European Concrete Committee are presented in a form which enables the web reinforcement to be calculated as a proportion of the amount required to resist the full shear. The expression for members cracked in flexure may be rearranged as follows, for comparison with the ACI formula:

$$v_u = \frac{V_u}{b_w d} = \frac{\zeta \rho_v f_y}{1.15} + 12.5\sqrt{f_c'} \left(1 + \frac{2f_{cp}}{f_c'}\right) + \frac{P \sin \theta}{b_w d},$$

where

ρ_v = ratio of shear reinforcement,

ζ = a coefficient, defined in Chapter 2 (p. 54), equal to 1 for reinforcement inclined at 45° or 90°.

f_y = yield stress of shear reinforcement,

f_c' = compressive strength of concrete (may be assumed 0.8 times cube strength),

f_{cp} = compressive prestress at centroid of concrete section,

ψ = coefficient which allows for the reduced shear strength of members with small amounts of tensile reinforcement near the support. It has the value $0.5 + 33 A_s/b_w d$ when $A_s/b_w d < 1.5\%$, otherwise it is equal to unity.

In this expression the contribution of the concrete increases with the prestress. It will be noted that the shear component of the prestressing force, $P \sin \theta$, is taken into account, whereas it is disregarded in the ACI method of calculation for flexure-shear cracking.

The above formula is also recommended for reinforced concrete members subject to compression and bending and a parallel formula has been proposed for members subject to tension and bending. In the latter the bracketed term $(1 + 2f_{cp}/f_c')$ is replaced by $(1 - 6f_c/f_c')$, where f_c is the average tensile

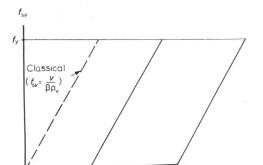

Figure 5.7   Stress in shear reinforcement

stress in the reinforced cross section under the combined effect of the normal external tensile force and the longitudinal prestress.

The method of calculation given by the European Concrete Committee for the shear reinforcement of prestressed members with web-shear cracking is based on the classical formula derived in Chapter 2 (p. 53f), but instead of the reinforcement being considered to replace the full principal tensile resistance of the concrete, the principal tensile stress is reduced by an amount $0.16\sqrt{f_c'}$ N/mm^2 to allow for the contribution of the concrete. The value of the principal tensile stress corresponding to the ultimate shear resistance is therefore given by

$$f_{max} = \rho_v f_y \frac{\sin \alpha \, \sin(\alpha - \phi)}{\cos \phi} + 0.16\sqrt{f_c'}$$

The shear resistance may be calculated by inserting this value in the classical equation, or the approximate version in the British Code, quoted above.

Some web reinforcement is necessary as it is possible for a very lightly reinforced member to fail suddenly by diagonal tension. The revised American Code recommendation for web reinforcement [5.1] is

$$\frac{A_{sv}}{b_w s} > \frac{50}{f_{yv}} \qquad (f_{yv} = \text{yield stress of shear reinforcement in lbf/in}^2).$$

The minimum requirement in the British Code is

$$\frac{A_{sv}}{b_w s} \geqslant 0.0012 \qquad \text{for high yield stirrups}$$

and

$$\frac{A_{sv}}{b_w s} \geqslant 0.002 \qquad \text{for mild steel stirrups}$$

For prestressed members the American Code specifies the following minimum

$$\frac{A_{sv}}{b_w s} \geqslant \frac{1}{80} \frac{A_p}{b_w d} \frac{f_{pu}}{f_{yv}} \sqrt{\frac{d}{b_w}} \ .$$

The corresponding expression in the British Code is:

$$\frac{A_{sv}}{b_w s} \geqslant \frac{0.4}{0.087 f_{yv}} \quad (f_{yv} = \text{yield stress of shear reinforcement in N/mm}^2).$$

The proposal of the European Concrete Committee is based on the amount of shear reinforcement required to develop a principal compressive stress of at least one quarter of the tensile strength of the concrete in any part of the web. When the web reinforcement bars are parallel and all of the same type this leads to the expression

$$\rho_v > \frac{f_{ct}}{4 f_{yv}} \ .$$

In addition the relative volume of the web reinforcement should not be less than 0.25% for smooth mild steel bars or 0.14% for high-quality high-bond steel. Wide beams, lintols and slabs are exceptions to the above requirements.

A further consideration in determining the shear strength of members is that beyond a certain level of web reinforcement failure of the concrete may occur. This may be in two possible ways. On the one hand the compressive zone may fail before the web reinforcement has reached its yield point. Alternatively, particularly in thin-webbed members, the diagonal concrete 'struts' formed between the inclined cracks may fail in compression as illustrated in Fig. 5.8. In

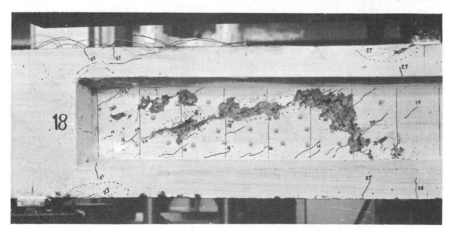

Figure 5.8   Shear failure of a prestressed beam by inclined compression of the concrete

practice both contingencies can be covered by restricting the average shear stress V/b_wd to a maximum value which is a function of the strength of the concrete, irrespective of the amount of reinforcement. In the British Code this maximum value is $0.75\sqrt{f_{cu}}$ N/mm^2, whereas the American Code specifies a maximum value of $5\sqrt{f_c'}$ lbf/in^2 $(0.42\sqrt{f_c'}$ N/mm$^2)$ for the concrete and a further amount of $8\sqrt{f_c'}$ lbf/in^2 $(0.66\sqrt{f_c'}$ N/mm$^2)$ for the reinforcement. The European Concrete Committee has recommended the following condition governing the principal compressive stress in the concrete:

$$-f_{min} \leqslant f_c' - 4 f_{max}$$

where

$$f_{max} \leqslant 0.125 f_c',$$

$$-f_{min} \leqslant 0.0625 \frac{f_c'}{f_{max}} \cdot f_c'$$

where

$$f_{max} \geqslant 0.125 f_c'.$$

When there is no transverse prestressing, the above inequalities may be replaced by

$$v \leqslant 0.25 f_c',$$

where v is the shear stress on the cross section of the beam at the level of the centroid.

Web reinforcement, to be effective, must be arranged so that each stirrup or bent-up bar is intersected by at least one inclined crack. This implies a maximum spacing of about $0.75\,d$ for vertical stirrups. Bent-up bars should be properly anchored and only the straight portion should be considered effective.

Cracking of members in pure torsion

The calculation of the shear stress due to the torsion of an uncracked member has been discussed in Chapter 1 (p. 31ff). In laboratory tests on reinforced concrete members in pure torsion, cracks have been found to appear when the calculated shear stress, which is equivalent to the principal tensile stress, is equal to about $6\sqrt{f_c'}$ lbf/in^2 (approx $0.45 f_{cu}$ N/mm^2) [5.9]. In the absence of special torsion reinforcement this must be taken to represent the strength of the member. The American Code proposals [5.9] require the provision of torsion reinforcement when the shear stress exceeds $2.4\sqrt{f_c'}$ lbf/in^2 (approximately $0.18f_{cu}$ N/mm^2). This amounts to only about 40% of the stress at cracking in pure torsion and has been reduced to allow for the effect of a bending moment acting together with the torque, as well as providing added safety against cracking. Less data is available on the torsional cracking load of prestressed members. A possible procedure is to calculate the torque at which the principal tensile stress reaches the above value, using the methods described in Chapter 1.

Cracking of members in combined shear and torsion

Shearing forces and torsion both give rise to shear and principal tensile stresses, and must therefore be considered together in determining the cracking load. Since the maximum shear stresses due to flexural shear and torsion do not necessarily occur at the same point, simple addition of the stresses is invalid and experiments have been carried out to investigate the combined effect in reinforced concrete beams of rectangular, T and L section without web reinforcement. Tests reported by Mattock [4.10] have indicated a circular interaction curve as shown in Fig. 5.9 between the variables V_u/V_o and T_u/T_o where V_u and T_u are the ultimate values of shear and torsion acting together and V_o and T_o the ultimate values of each alone. The interaction equation is therefore

$$\left(\frac{V_u}{V_o}\right)^2 + \left(\frac{T_u}{T_o}\right)^2 = 1.$$

This may be written in terms of the normal shear stresses, v_v due to shearing force and v_t due to torsion, so that

$$\left(\frac{v_{vu}}{v_{vo}}\right)^2 + \left(\frac{v_{tu}}{v_{to}}\right)^2 = 1$$

where

$$v_v = \frac{V}{b_w d}, \quad v_t = \frac{3T}{\Sigma h_{min}^2 h_{max}}$$

h_{min} and h_{max} are the shorter and longer sides of the component rectangles of the section. Therefore

$$v_{tu} = \frac{v_{to}}{\sqrt{\left\{1 + \left(\frac{v_{vu} v_{to}}{v_{tu} v_{vo}}\right)^2\right\}}}.$$

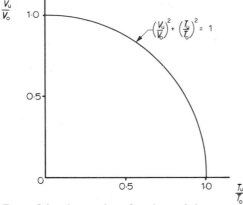

Figure 5.9   Interaction of torsion and shear

The values of v_{vo} and v_{to} at which cracking occurs have been found by experiment to be approximately $2\sqrt{f_c'}$ and $6\sqrt{f_c'}$ respectively. Therefore if it is assumed that the ratio v_v/v_t remains constant throughout the loading history of the member,

$$v_{tu} = \frac{v_{to}}{\sqrt{1 + \left(\dfrac{3v_{vu}}{v_{tu}}\right)^2}} \quad .$$

Similarly

$$v_{vu} = \frac{v_{vo}}{\sqrt{1 + \left(\dfrac{v_{tu}}{3v_{vu}}\right)^2}} \quad .$$

Strength of members with torsion reinforcement

The ideal orientation of torsion reinforcement would be in a spiral so as to intersect the torsion cracks at approximately right angles, but for practical reasons it usually consists of a system of closed stirrups.

The action may be considered to be that of a truss, analogous to the action of shear reinforcement, with the concrete functioning as compressive members, and it will be seen from Fig. 5.10 that there are horizontal as well as vertical forces, requiring that each stirrup should be completely closed. Two further points that emerge from Fig. 5.10 are that a tensile force is taken up by the longitudinal corner bars and that the spacing of the stirrups must be limited to allow each crack to be intersected by at least one stirrup on each face of the member.

The maximum spacing of the stirrups recommended is x_1 and $0.5\,y_1$ or 200 mm in the British Code, where x_1 and y_1, are respectively the shorter and longer centre to centre dimension of the stirrups. In the American Code proposals [5.9] the corresponding spacings are $1.33\,x_1$ and $0.5\,y_1$. The torsional strength is improved when there are longitudinal bars not only at the corners, but also distributed as evenly as possible around the perimeter of the stirrups, and a maximum spacing of $1.33\,x_1$ or 460 mm (18 in) has been recommended in the American Code, and 300 mm in the British Code.

When members of T, L, I, or box section have to be reinforced against torsion it is necessary that each component rectangle of the section should contain its own rectangular stirrups, preferably closed, extending into the adjacent part of the section.

The general form of the expression for the ultimate torsional resistance of stirrups is

$$T = \psi \frac{x_1 y_1 A_{sv} f_y}{s}$$

where

A_{sv} = cross sectional area of two legs of stirrup,

x_1 = shorter centre to centre dimension of stirrups,

y_1 = longer centre to centre dimension of stirrups,

f_y = yield stress of stirrups,

ψ = constant

Various values have been recommended for the constant ψ. Rausch [5.11] originally derived the formula from the circular section theory of torsion for

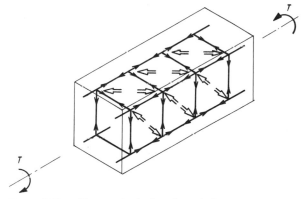

Figure 5.10   'Truss action' of torsion reinforcement

which $\psi = 1.0$. Andersen [5.12] obtained $\psi = 0.67$ by assuming a parabolic stress distribution in the reinforcement and by deriving the radius of the equivalent circular section by the Bach formula. Cowan [5.13] used a strain energy method and showed ψ to lie between 0.80 and 0.85 where the ratio of the sides of the section did not exceed 3; the lower value was recommended for design. More recently the American Code proposal [5.9] has given $\psi = 0.33 + 0.16\ (y_1/x_1)$, with a maximum of 0.75, on the basis of experimental work by Hsu [5.14]. The British Code proposes a value of 0.80 and the European Concrete Committee 1.00; in the latter recommendations for a box section, x_1 and y_1 are measured between the centre lines of the walls.

Since under conditions of pure torsion, the stirrups oppose a principal tensile stress inclined at $45°$ to the axis, an equal volume of complementary longitudinal reinforcement is required. The area A_{sl} of this reinforcement is therefore given by

$$A_{sl} = A_{sv} \left(\frac{x_1 + y_1}{s} \right).$$

The American Code proposals suggest a contribution of the concrete to the torsional strength corresponding to a nominal shear stress of $2.4\sqrt{f_c'}$ lbf/in^2 (approximately $0.18\sqrt{f_{cu}}$ N/mm^2), the same value as the allowance for members without torsion reinforcement and analogous to the method of calculation for flexural shear. This is reduced when combined with the flexural shear according to the circular relationship described above. No corresponding allowance is made in the European Concrete Committee and British Code proposals and it is considered that in a large member with fully developed spiral torsion cracks the only effective resistance is that of the reinforcement.

An upper limit to the torsional strength, independent of the amount of torsion reinforcement, is imposed by the diagonal compression strength of the concrete in the same manner as in flexural shear. In America [5.9] the maximum nominal shear stress has been empirically fixed as $14\sqrt{f_c'}$ lbf/in^2 (approximately $1.04\sqrt{f_{cu}}$ N/mm^2).

The criterion for the maximum torsional strength proposed by the European Concrete Committee is given by the formula

$$v_t < \frac{T}{2A_o b_o}$$

where

A_o = Area of the section taken halfway through the walls, or $x_1 y_1$ in a rectangular section in which the lateral spacing of the longitudinal corner bars is less than $5b/6$,

b_o = Thickness of walls in a box section.

In a rectangular section the equivalent wall thickness is assumed to be $b/6$.

The value of the stress v_t corresponding to diagonal compression failure of the concrete is assumed to be $0.18 f_c'$ (approximately $0.14 f_{cu}$) or 4.5 N/mm^2 whichever is less, when the torsion reinforcement consists of orthogonal stirrups. The stress is assumed to be $0.22 f_c'$ (approximately $0.18 f_{cu}$) or 5.5 N/mm^2 for steel reinforcement at $45°$ to the axis.

The British Code gives approximate formulae based on a plastic distribution of the shear stress due to torsion.

For rectangular sections

$$v_t = \frac{2T}{h_{min}\left(h_{max} - \dfrac{h_{min}}{3}\right)}$$

where h_{min} and h_{max} are the shorter and longer dimensions of the section.

It is recommended that T , L , or I sections should be treated by dividing them into their component rectangles in such a way as to maximise $\Sigma(h_{min}^3 h_{max})$, which can generally be done by making the widest rectangle as long as possible. The torsional shear stress on each rectangle is then calculated by treating it as a separate rectangular section subjected to a torsional moment of

$$\frac{T(h_{min}^3\, h_{max})}{\Sigma(h_{min}^3\, h_{max})}.$$

Box sections in which the wall thickness exceeds one quarter of the overall dimensions of the member in that direction, are to be treated as rectangular sections.

When members are subject to both torsional and flexural shear the above maximum values have to be reduced. In the American Code the circular interaction curve is assumed and it will be recalled that the maximum shear stress due to flexural shear is $10\sqrt{f_c'}$ lbf/in^2 (approximately $0.75\sqrt{f_{cu}}$ N/mm^2). Hence v_{tu}, the ultimate torsional shear stress when accompanied by a flexural shear stress v_{vu}, is given by

$$v_{tu} < \frac{14\sqrt{f_c'}}{\sqrt{\{1 + (1.4\, v_{vu}/v_{tu})^2\}}}$$

The European Concrete Committee has adopted a linear interaction relationship

$$\frac{v_{vu}}{v_{vo}} + \frac{v_{tu}}{v_{to}} < 1,$$

where v_{vo} is the maximum flexural shear stress in the absence of torsion and the maximum torsional shear stress in the absence of flexural shear, specified as described earlier.

In the British Code the maximum allowable value of the sum of the shear stresses due to flexural shear and torsion acting together is between 3.35 and 4.75 N/mm^2, depending on the grade of concrete.

Lateral instability of columns and second-order effects

The classical theory of lateral elastic instability of slender columns originates from the work of Euler, who showed that the axial load at which a column, hinged but fixed in position at the ends, becomes unstable (i.e. buckles) is

$$N_e = \frac{\pi^2 EI}{2l}$$

where l is the height of the column and EI the flexural rigidity. The corresponding 'Euler' stress is therefore

$$f_e = \frac{\pi^2 E}{(l/i)^2}$$

where i is the radius of gyration of the section about the axis of bending which gives i its least value.

Various other conditions of end restraint are accommodated by the use of a modified value of l, known as the effective length.

This theory requires some adjustment, even for axial loads on hinged columns, when applied to reinforced concrete, because of the changes in the flexural rigidity brought about by cracking and by non-linear deformation of the concrete and steel at high stresses. When a column is loaded eccentrically, however, the lateral deflexion due to flexure results in the eccentricity of the load varying along the length, so that the maximum moment may be greater than the values at the ends. This so-called 'second-order' effect cannot be separated from the problem of lateral stability.

Von Karman [5.15] developed a general analytical method for determining the strength of an eccentrically loaded column in which the maximum stresses exceeded the limit of porportionality of the material. The rotation angles of each short length were found from the stress-strain relationship, assuming the strain distribution to be linear at each section, and the deflected column profile was obtained by numerical integration. The process was repeated for an increasing load until the total moment, including the second-order moment, exceeded the flexural strength at the point of maximum moment.

Baumann [5.16] applied von Karman's method to reinforced concrete and found good agreement with the available results of tests. The numerical integration procedure was tedious, however, and until about 1960 calculations usually introduced the simplifying assumption that the deflected column profile could be represented by a cosine wave [5.17, 5.18, 5.19]. It then became possible to rapidly perform numerical integration and iteration procedures by electronic digital computer, and the important contributions of Chang and Ferguson [5.20] and of Pfrang and Siess [5.21, 5.22] marked a return to the more fundamental approach of von Karman.

In recent years many types of slender column with various degrees of fixity and loading conditions have been analysed by numerical integration methods [5.20, 5.22, 5.23, 5.24, 5.25] and a good number of the results validated by laboratory tests [5.20, 5.24, 5.25]. In the effort to establish a reasonably simple and accurate method for design, two main alternatives have emerged.

The first and more traditional of these two alternatives uses the form of the Euler equation, that is, the ultimate load is considered to decrease with an increasing slenderness ratio l/i, so that the ultimate load obtained for a short column must be multiplied by a reduction factor. This is the method given in British Codes up to the present unified code, and in the 1963 American Code; the reduction factors recommended are illustrated in Fig. 5.11 together with those obtained by the Euler formula for typical values of the strength and modulus of elasticity of concrete.

In an eccentrically loaded column the above reduction factors, applied to the load, reduce the moment in the same proportion. The effect is thus to reduce the scale of both axes of the load-moment curve, as in Fig. 5.12. The alternative

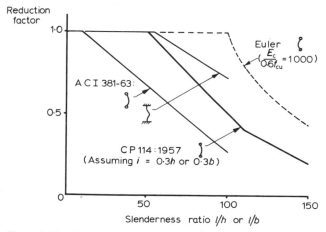

Figure 5.11   Load reduction factors for slender columns

approach, advocated by the European Concrete Committee and introduced into the Codes of several countries is to consider the effect of slenderness as equivalent to an increase in the eccentricity of the load, so that the proportional reduction of the moment capacity of a slender column, as compared with that of a short column, is the same for all values of the load. This type of modification to the load-moment curve is also shown in Fig. 5.12.

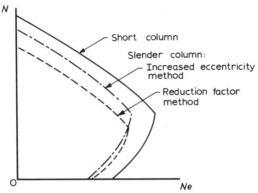

Figure 5.12   Load-moment characteristic of a slender column

The European Concrete Committee has developed the following approximate empirical expression for the supplementary moment ΔM due to the second-order effects

$$\Delta M = \frac{Nl^2}{10r}$$

where $1/r$ = curvature of member in the centre of the buckling length, given as follows:

(a)   For $N \leqslant 0.5 f_c' A_c$,

$$\frac{1}{r_1} = \frac{(-\epsilon_c + \epsilon_s)}{h} - \frac{l}{50\,000d^2} ,$$

where the strains in the steel and concrete are

$$\epsilon_c = 0.0003 \text{ and } \epsilon_s = \frac{f_y}{E_s}$$

(b)   For $N \geqslant 0.5 f_c' A_c$,

$$\frac{1}{r_2} = \frac{f_c' A_c}{2r_1 N} .$$

For design, the strengths of the steel and concrete are reduced by partial factors as discussed in Chapter 6. In addition, in order to allow for creep when considering permanent loads, the value of ϵ_c is multiplied by $(1 + \psi\beta)$ where β is the ratio of permanent load to total load and ψ is taken from the following table:

Period at which all of the permanent load is applied,	ψ	
	Atmospheric conditions	
	temperate or humid	dry
(1) at the end of the construction time	0.5	1.0
(2) six months (or more) after completion of the works	0.25	0.5

In the British Code recommendations, the stability is considered about both axes of the section. The expression for the supplementary moment, in columns of rectangular section having a symmetrical arrangement of reinforcement, is of the form

$$\Delta M = \frac{Nh}{1750} \, (l/h)^2 \, (1 - 0.0035 \, l/h).$$

Studies in America [5.23, 5.24, 5.25] have compared the unfactored strengths, given by the (A.C.I. 1963) reduction factor method and the European Concrete Committee (C.E.B. 1964) additional moment method, with values obtained by numerical analysis. It was found that where there was no lateral displacement of the end, the A.C.I. method was conservative for columns deformed in single curvature but sometimes unsafe for double curvature, while the opposite was true of the C.E.B. method. The latter gave a better indication of the mode of failure of a slender column in single curvature, but neither method determined the position of the failure section.

Although creep under sustained load has comparatively little effect on the strength of concrete sections, the increase of lateral deflexion can cause a more important reduction in the strength of a column. In one frame test with a particularly unfavourable lateral load a column collapsed at only 60% of the short-term ultimate load [5.26].

Lateral instability of reinforced concrete beams

Lateral instability may arise in beams having too great a length between lateral restraints in relation to the lateral and torsional stiffness. The theoretical expression for the critical moment M_{crit} which will just cause instability in a homogeneous beam with a linear stress-strain characteristic, as derived by Michell [5.27] is

$$M_{crit} = \frac{\beta}{l} \sqrt{\left(\frac{EI_x EI_y GJ}{EI_x - EI_y}\right)} \left\{1 - \frac{1.74y}{l} \sqrt{\left(\frac{EI_y}{GJ}\right)}\right\},$$

where

β = coefficient depending on type of loading,

l = distance between lateral restraints,

EI_x = flexural rigidity about major axis,

EI_y = flexural rigidity about minor axis,

GJ = torsional rigidity about axis through centroid,

a = distance of point of application of load above centroid of section.

If EI_y is small in relation to EI_x this may be simplified to

$$M_{crit} = \frac{\beta}{l} \sqrt{(EI_y GJ)} \left\{1 - \frac{1.74a}{l} \sqrt{\left(\frac{EI_y}{GJ}\right)}\right\}.$$

The problem of applying this expression to the lateral instability of reinforced concrete beams, in which the values of the flexural and torsional rigidities are dependent on the load, has recently been examined by Marshall [5.28]. The criterion of whether an unsupported length of beam will be unstable is found to be the parameter ld/b^2 where d is the effective depth and b the breadth of the compression flange. By assessing the extreme values of the different variables it was shown that, for instability to occur before flexural failure, ld/b^2 might be expected to lie between 104 and 568. Experimental results usually indicated a value equal to, or greater than, the higher of these limits, suggesting an ultimate load criterion of $ld/b^2 \simeq 500$ and $M_{crit} \simeq 140(f_{cu} b^3 d)/l$ under a uniformly distributed load.

The ratio l/d is commonly used in codes of practice as a very approximate guide for lateral stability. A maximum value of 50 is specified in the American Codes, where b is the breadth of the compression flange and l the distance between lateral supports.

Lateral instability of prestressed beams during erection

A particular problem may occur during the lifting of slender prestressed beams, where tilting about the longitudinal axis results in lateral bending, the effect of which may be aggravated if the prestress acts in the same direction as the bending stresses. Additional reinforcement is then advisable, particularly at the corners of the compression flange.

REFERENCES

[5.1] McGregor, J. G. and Hanson, J. M. (April 1969) proposed changes in shear provisions for reinforced and prestressed concrete beams. *Proc. Amer. Concr. Inst.*, **66**, 276-288.

[5.2] Shear study Group of Institution of Structural Engineers. (Jan. 1969) The shear strength of reinforced concrete beams. *Instn. Struct. Engnrs.*, Series No. 49.

[5.3] ACI Committee 318. (1965) Commentary on Building Code requirements for reinforced concrete. Publn. SP-10 Detroit Amer. Concr. Inst.

[5.4] ACI-ASCE Committee 326 (Jan.–Feb. 1962) Shear and diagonal tension. *Proc. Amer. Concr. Inst.*, **59**, 1-30, 277-334.

[5.5] Mattock, A. H. (August 1968) Diagonal tension cracking in concrete beams with axial forces. *Dept. Civil Eng., Struct. Res.* Report 68-3. Seattle, Washington: University of Washington.

[5.6] Ferguson, P. M. (August 1965) Some implications of recent diagonal tension tests. *Proc. Amer. Concr. Inst.*, **53**, 157-172.

[5.7] Leonhardt, F. (Dec. 1965) Reducing the shear reinforcement in reinforced beams and slabs. *Mag. Concr. Res.*, **17**, 183-198.

[5.8] Leonhardt, F. (1965) *Schubbemessung bei Spannbetontragwerken* (*Shear measurements in prestressed concrete beam structures*). Berlin: Deutscher Betontag.

[5.9] ACI Committee 438. (Jan 1969) Tentative recommendations for the design of reinforced concrete members to resist torsion. *Proc. Amer. Concr. Inst.*, **66**, 1-8.

[5.10] Mattock, A. H. (1968) How to design for torsion. In *Torsion of structural concrete*, SP-18, (pp. 469-495.) Detroit: Am. Concr. Inst.

[5.11] Rausch, E. (1929) *Berechnung des Eisenbetons gegen Verdrehung und Abscheren* (*Design of reinforced concrete for torsion and shear*). Berlin: Springer Verlag.

[5.12] Andersen, P. (Sept.–Oct. 1938) Rectangular concrete section under torsion. *Proc. Amer. Concr. Inst.* **34**, 1–11.

[5.13] Cowan, H. J. (July 1950) An elastic theory for the torsional strength of rectangular reinforced concrete beams. *Mag. Concr. Res.* **2**, *No. 4*, 3-8.

[5.14] Hsu, T. C. C. (1968) Torsion of structural concrete – behaviour of reinforced concrete rectangular members. In *Torsion of structural concrete*, SP-18. Detroit: Amer. Concr. Inst.

[5.15] Karman, T. von (1910) Untersuchungen uber Knickfestigkeit (Investigation of instability). Mitteilungen uber Forschungsarbeiten auf dem Gebiete des Ingenieurwesens, No. 81, Berlin.

[5.16] Baumann, O. (1934) Die Knickung der Eisenbeton–Saulen (The buckling of reinforced concrete columns). Eidg. Material prufungsanstalt an der E.T.H. in Zurich, Bericht, No. 89, Zurich.

[5.17] Broms, B. and Viest, I. M. (Jan. 1958) Ultimate strength analysis of long hinged reinforcement concrete columns. *Proc. Amer. Soc. Civ. Engnrs., 84, No. ST1*, 1510-1 to 1510-38.

[5.18] Broms, B. and Viest, I. M. (May 1958). Ultimate strength analysis of long restrained reinforced concrete columns. Proc. Am. Soc. Civ. Engnrs. 84, No. ST3, pp. 1635-1 to 1635-30.

[5.19] Broms, B. and Viest, I. M. (July 1958) Design of long reinforced concrete columns. *Proc. Amer. Soc. Civil Engnrs., 84, No. ST4*, 1694-1 to 1694-28.

[5.20] Chang, W. F. and Ferguson, P. M. (Jan. 1963) Long hinged reinforced concrete columns. *Proc. Amer. Concr. Inst., 60*, 1-25.

[5.21] Pfrang, E. O. and Siess, C. P. (Oct. 1964) Predicting structural behaviour analytically. *Proc. Amer. Soc. Civil Engnrs., 90, No. ST5*, 99-112.

[5.22] Pfrang, E. O. and Siess, C. P. (Oct. 1964) Behaviour of restrained reinforced concrete columns. *Proc. Am. Soc. Civil Engnrs, 90, No. ST5*, 113-136.

[5.23] Martin, I., MacGregor, J. G., Pfrang, E. O. & Breen, J. E. (1966) Critical review of the design of reinforced concrete columns. *Symp. Reinf. Concr. Columns*, Publn. SP-13. Detroit: Amer. Concr. Inst.

[5.24] Martin, I., and Olivieri, E. (1966) Test of slender reinforced columns bend in double curvature. *Symp. Reinf. Concr. Columns*, Publn. SP-13. Detroit: Amer. Concr. Inst.

[5.25] MacGregor, J. G., and Barter, S. L. (1966) Long eccentrically loaded concrete columns bent in double curvature. *Symp. Reinf. Concr. Columns*, Publn. SP-13. Detroit: Amer. Concr. Inst.

[5.26] Ferguson, P. M. and Breen, J. E. (1966) Investigation of the long concrete column in a frame subject to lateral loads. *Symp. Reinf. Concr. Columns*, Publn. SP-13. Detroit: Amer. Concr. Inst.

[5.27] Michell, A. G. M. (1899) The elastic stability of long beams under transverse forces. *Phil. Mag., 48, No. 292.*

[5.28] Marshall, W. T. (July 1969) A survey of the problems of lateral instability in reinforced concrete beams. *Proc. Instn. Civil Engrs., 43*, pp. 397-406.

Part II
Design of Structural
Concrete Elements

6

Principles of structural design

Design requirements — general and structural

The preceding chapters have been concerned with the different modes of behaviour of various types of structural concrete member, according to the nature and degree of loading. The examination of this behaviour, aided by the principles of structural mechanics, has led to methods of calculation of the strength and deformation of such elements. Design, however, is essentially a creative rather than an analytical activity, in which the structural behaviour is only one of a number of functional, constructional, aesthetic and economic considerations. The dimensioning of structural concrete elements cannot therefore be completely isolated from this broader context, and it is unrealistic to attempt to set up a rigid system whereby a given set of loading conditions and material properties invariably results in a certain type and size of member, nor will the use of the smallest sections or minimum quantities of steel always yield the most economical solution. Therefore, in examining how the required dimensions of a member are affected by loading requirements and strength specifications, it is necessary to provide ample flexibility for these other considerations to be taken into account.

The specifically structural requirements in design are comprehended in the broad categories of safety and serviceability. Safety may be defined as an acceptable degree of security against complete failure, which in concrete structures can occur by the various modes discussed in Chapters 4 and 5. The requirement of serviceability means that the member or structure should not in its intended lifetime deteriorate to such an extent that it no longer fulfils the function for which it was designed. In structural concrete this may arise through defective structural behaviour such as excessive deflexion or wide cracks, or by chemical attack of the concrete and steel in a corrosive environment.

Permissible stress method of design

Until recently two methods have been in general use for the design of structural concrete members: the permissible stress method and the load factor method. These have sometimes been used as alternatives and sometimes used in combination.

The permissible stress method was used in England prior to 1957, for all reinforced concrete members with the exception of columns. In the subsequent Code of Practice (CP 114: 1957) it was retained as an alternative to the load factor method for flexure, with and without compression, and remained the basis of design for shear and bond. In the first Code for prestressed concrete (CP 115: 1959) both permissible stress and load factor conditions were jointly imposed.

In the permissible stress method the calculations relate to a working load which, it is assumed, will not be exceeded in the normal life of the structure. The stresses corresponding to this level of load must not exceed specified permissible values, appropriate to the material. For example, in the above code the permissible tensile stress in steel reinforcement ($f_{s\,adm}$) was normally the yield stress divided by the factor 1.8, while the permissible compressive stress for concrete in bending ($f_{c\,adm}$) was the works cube strength divided by the factor 3. The permissible stress was sometimes determined by the criterion of serviceability rather than strength; for example the maximum permissible steel stress was 230 N/mm^2 (33 000 lbf/in^2) to avoid an undesirable degree of cracking and deflexion.

The calculations are made by the methods described in Chapter 2, assuming a linear stress-strain relationship and a constant modular ratio of steel to concrete. Since the ratio of the steel stress to the concrete stress is determined by the dimensions of the section and reinforcement, the permissible stress will generally be attained in only one of the two materials, the stress in the other being less than the permissible value. This is seen in the example on page 38ff. Both materials will attain their permissible stresses simultaneously in a 'balanced section' with a critical value of the reinforcement ratio, sometimes incorrectly termed the 'economic percentage.'

Load factor method of design

The main objection to the permissible stress method is that the stress safety factor, relating the stress permitted at design load to the strength of each material, is not usually the same as the ratio of the strength to the design load for the section, member or structure as a whole, so that the method does not give an adequate indication of the degree of safety achieved. This is because the system of internal stresses at the point of failure usually differs from that which occurs at the design load, as will readily be seen from a comparison of corresponding stress distribution diagrams in Chapters 2 and 4. The load factor method is intended to overcome this disadvantage by the substitution of a strength calculation, using the methods discussed in Chapters 4 and 5, for a calculation of the resistance corresponding to the permissible stresses. The degree of safety is then controlled by the magnitude of the ratio strength/ working load, known as the load factor.

The 1957 code for reinforced concrete (1967 reprint) permitted the use of this method with a central load factor of 1.8. However, in order to allow for the fact that there is a greater variation in the strength of concrete than in the strength of steel, it was specified that in calculating the strength, the cube strength of concrete should first be reduced in the ratio 1.8/3 for nominal mixes, or 1.8/2.73 for designed mixes. In order to bring the load factor method into line with the alternative permissible stress method, the various strength formulae were converted into resistance formulae at design load level, by the insertion of the factored strengths, yield stress/1.8 for steel and works cube strength/3 for concrete. These factored strengths were identical to the permissible stresses used in the alternative method. A somewhat similar concept whereby the design strength of a member is based on an appropriate partial factor applied to the strength of each constituent material has been adopted by the European Concrete Committee and also used in the new British Code.

In the Prestressed Concrete Code (1959) a strength calculation was demanded, the central load factor having a maximum value of 2. A further innovation was the use of different partial factors for the dead load and the live load components of the total working load, with the values 1.5 and 2.5 respectively. The larger factor for the live load was specified to take account of the greater uncertainty and the possibility of an accidental overload. The design condition was therefore

$$1.5\,F_g + 2.5\,F_q \leqslant R_u$$

or

$$2.0\,(F_g + F_q) \leqslant R_u, \quad \text{if } F_q > F_g$$

where

$$F_g = \text{dead load},$$
$$F_q = \text{live load},$$
$$R_u = \text{calculated ultimate resistance}.$$

The American code contains a similar type of condition with additional requirements where wind and earthquake loadings have to be considered. A further feature in this code is that the calculated ultimate resistance is multiplied by a coefficient ϕ, less than unity, so that the design condition for dead and live load, given above, becomes

$$1.4F_g + 1.7F_q \leqslant \phi R_u.$$

The coefficient ϕ, known as the capacity reduction factor, is provided to allow for possible reduction of the strength due to variation in the quality of the materials, inaccuracy in construction and the degree of approximation inherent in the formula used to calculate the ultimate resistance. The specified values of ϕ are

Flexure 0.90
Diagonal tension, bond and anchorage 0.85
Spirally reinforced compression members 0.75
Tied compression members .... 0.70.

Variability of load

It is now necessary to consider the magnitude of the load factor in relation to the actual safety of the member. The degree of safety is considerably affected by the variation to which both load and ultimate resistance may be subject, and accuracy in the assessment of safety therefore depends on a knowledge of the probable extent of this variation. The load factor method can, in fact, only be regarded as satisfactory when proper account is taken of the variability of load and resistance.

The maximum loads occurring during the lifetime of each of a number of structural members will actually be different, even though the loading conditions are nominally the same. They may conveniently be represented by the wellknown block diagram or histogram, an example of which is shown in Fig. 6.1, in which the number of occurrences of a load in each interval of magnitude on the horizontal scale is denoted by the vertical scale. If the width of each interval of load were to be made infinitesimal the histogram would be transformed from a series of orthogonal lines to a smooth curve, in which the area enclosed between any two vertical lines would correspond to the number of members having a maximum load of magnitude between the limits represented by the two lines. Furthermore, if the vertical scale were adjusted so that the total area bounded by the curve and the horizontal axis, instead of representing the total number of load occurrences, became equal to unity, the area enclosed between the two vertical lines would denote the fraction of all the members having a maximum load within the given limits of magnitude. This would be the same as the probability ψ of any member, chosen at random, having a maximum load within these limits.

The curve thus described is known as a distribution curve, and if the horizontal scale represents the load F, the units of the vertical scale will be $d\psi/dF$, known as probability density (Fig. 6.1).

Two characteristic features of a distribution curve are the mean value of F,

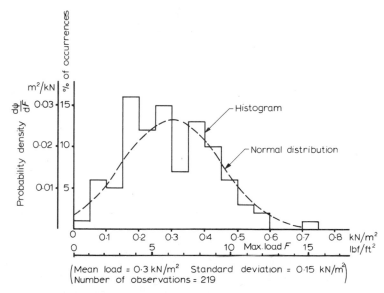

Figure 6.1   Maximum loads due to persons on floor structures of flats over a period of 10 years (A. I. Johnson)

i.e. the abscissa of the centroid of the area of the curve, and the amount of 'spread' or 'scatter' of the load occurrences on each side of this mean. The latter can be represented by the radius of gyration of the area of the curve about a vertical axis through the centroid, which is known as the standard deviation. The mean F_m and standard deviation s_F may be obtained approximately from a finite number n of data as follows:

$$F_m = \frac{\Sigma F}{n},$$

$$s_F = \frac{\Sigma (F - F_m)^2}{n - 1}.$$

Various mathematical expressions have been developed to represent distribution curves for different conditions. These enable the probability of the occurrence of values within certain limits to be estimated, particularly the probability of a value lying above or below a chosen limit. The best known of these mathematical curves is the Normal or Gaussian distribution, (Fig. 6.1) represented by the formula

$$\frac{d\psi}{dF} = \frac{1}{s_F \sqrt{(2\pi)}} e^{-\frac{1}{2}(F - F_m/s_F)^2}.$$

The probability of any individual load exceeding a value F_k is therefore

$$\psi = \frac{1}{s_F \sqrt{(2\pi)}} \int_{F_k}^{\infty} e^{-\frac{1}{2}(F - F_m/s_F)^2} \, dF.$$

Substituting $\nu = \dfrac{F - F_m}{s_F}$ this becomes

$$\psi = \frac{1}{\sqrt{(2\pi)}} \int_{\nu F_k}^{\infty} e^{-\frac{1}{2}\nu^2} \, d\nu.$$

Statistical tables have been compiled for this function; these enable the value of ν_{Fk} to be obtained for any desired level of probability. For example the value of ν_{Fk} for a probability of $2\frac{1}{2}\%$ is 1.95; that is to say, one member in 40 might be expected to undergo a maximum load exceeding the mean maximum load by more than about twice the standard deviation. This, of course, assumes that the normal distribution is a valid representation of the frequency of load occurrence.

The working loads specified for design in the past, and used in both the permissible stress and the load factor methods of design, have usually been decided by the judgement of experienced engineers rather than on a statistical basis. However, as and when more observations of actual loads become available the method described above may be used to establish a working load from probabilistic considerations. A working load arrived at on this basis is generally termed a characteristic load and is given by the expression

$$F_k = F_m + \nu_{Fk} s_F$$

where ν is a coefficient depending on the probability of the maximum load on any member exceeding the characteristic load.

The European Concrete Committee proposes that the mean load F_m should be defined as the value of the most unfavourable loading, with a 50% probability of not being greatly exceeded in the expected life of the structure, i.e. the mean of the maximum loads in the life-time of a number of similar structures. In the example given in Fig. 6.1, for a 5% probability that the average maximum load will be exceeded in the life of the structure, $\nu_{Fk} = 1.64$ so that

$$F_k = 0.3 + 1.64 \times 0.15 = 0.54 \text{ kN/m}^2 \text{ (11 lbf/ft}^2).$$

Statistical load observations of this type are not generally available, however, and in their absence the nominal loads recommended in the various codes and standards may be treated as characteristic loads. The British Code has adopted the loads in the Code CP3, chapter 5, as characteristic loads.

Variability of resistance

The variability of the ultimate resistance must also be considered. Obviously the ultimate resistance of a particular member has a unique value, but this can only be ascertained by a test to destruction. If a number of nominally identical members are tested, the values of the ultimate resistance will be found to follow a distribution curve, as did the maximum loads experienced by a number of different members in their lifetime. Thus the ultimate resistance of the member in question, which cannot be tested, is unknown but subject to variation according to the laws of chance as represented by the distribution curve. A value, termed a characteristic resistance, may therefore be defined, below which only a specified proportion of the ultimate resistance values will fall; this proportion represents the probability of the unknown ultimate resistance of a particular member being less than the characteristic value.

If the normal distribution curve applies, we therefore have

$$R_k = R_m - \nu_{Rk} s_R$$

where

$$R_k = \text{characteristic resistance,}$$

$$R_m = \text{mean resistance,}$$

$$s_R = \text{standard deviation of resistance,}$$

$$\nu_{Rk} = \text{coefficient related to probability of resistance being less than } R_k.$$

The necessary data to establish R_m and s_R for a complete member are only available on rare occasions, such as, for example, when a large number of tests to destruction have been made on a standard precast concrete product. It is, however, easier to obtain information on the variation of the strength of the individual materials, steel and concrete, of which the member is constructed. Tests of a number of samples enable the mean strength and standard deviation and hence the characteristic strength of each material to be determined.

As an example, Fig. 6.2 shows the histogram of compressive strength tests made on concrete cubes [6.9]. The mean strength is 34.5 N/mm² (5030 lbf/in²) and the standard deviation 6.2 N/mm² (900 lbf/in²). The characteristic strength of a material is defined in the British Code as the value below which 5% of individual values may be expected to lie, i.e. the probability of the actual strength of the material in a member exceeding the characteristic strength is 95%. Therefore if the distribution is normal, the characteristic strength will be 1.64 times the standard deviation below the mean strength and in this example the characteristic strength of the concrete will be

$$34.5 - 1.64 \times 6.2 = 24.3 \text{ N/mm}^2 \text{ (3550 lbf/in}^2).$$

The mean ultimate resistance of a structural concrete member may be obtained by inserting the mean strength of the steel and concrete in the ultimate resistance formula, but the standard deviation of the ultimate resistance of a member controlled equally by the strength of the steel and the concrete is given approximately by

$$\frac{s_R}{R_m} = \sqrt{\left\{ \left(\frac{s_{Rs}}{R_{ms}} \right)^2 + \left(\frac{s_{Rc}}{R_{mc}} \right)^2 \right\}}$$

where

$$R_{ms} = \text{mean strength of steel,}$$

$$R_{mc} = \text{mean strength of concrete,}$$

$$s_{Rs} = \text{standard deviation of strength of steel,}$$

$$s_{Rc} = \text{standard deviation of strength of concrete.}$$

The dimensionless term s_R/R_m is known as the coefficient of variation.

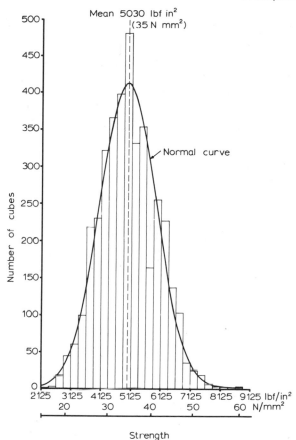

Strength

Figure 6.2   Histogram of compressive strength tests on concrete cubes (Erntroy)

Probability of failure

If the distribution curves are known for the maximum load, F, on a certain type of structural member during its lifetime, and for its resistance, R, (Fig. 6.3(a)) it is possible to calculate the probability that the member will fail at some time during its normal life period, i.e. the probability that F will be greater than R. This can best be done as suggested by Corso [6.1], by considering the distribution curve of their difference, $D = R - F$ (Fig. 6.3(b)), which is a closer approximation to the normal distribution than the distribution curves of R and F separately [6.2].

The probability of failure, ψ_u is the probability of D being negative, which is given by

$$\psi_u = \frac{1}{\sqrt{2\pi}} \int_{-\infty}^{0} e^{-\frac{1}{2}(D - D_m / s_D)^2} \frac{dD}{s_D}$$

where

$$D_m = R_m - F_m = \text{mean of } D,$$

$$s_D = \sqrt{(s_R{}^2 + s_F{}^2)} = \text{standard deviation of } D.$$

Making the substitution $v_D = (D_m - D)/s_D$, so that $dD = -s_D dv_D$ and the limits of the integral become D_m/s_D and $+\infty$,

$$\psi_u = \frac{1}{\sqrt{(2\pi)}} \int_{s_D}^{\infty} \frac{D_m}{s_D} e^{-\frac{1}{2}vD^2} \, dv_D.$$

ψ_u can therefore be obtained for the given value of D_m/s_D from a table of areas of the normal distribution function, to give the relationship shown in Fig. 6.4.

The probability of failure gives a quantitative indication of safety which is not provided by a load factor, and still less by a permissible stress. In attempting to assign an acceptable value to this probability, however, one becomes involved in philosophical questions as well as purely engineering considerations. One way

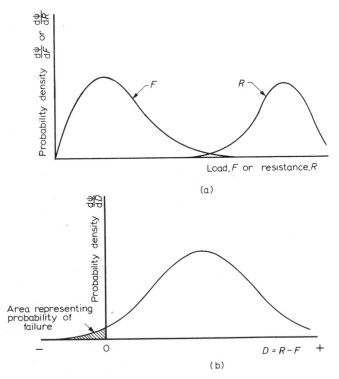

Figure 6.3   (a) Distribution curve of load F, and resistance R
(b) Distribution curve of R-F

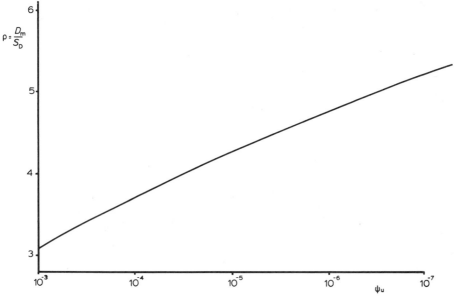

Figure 6.4   Relationship between reliability factor ρ and probability of failure ψ_u for normal distribution of R-F

in which the problem may be approached is by the concept of an accident rate intuitively accepted for a given type of structure. Pugsley [6.3] discusses this with particular reference to war-time experience of aircraft accidents in which a structural accident rate greater than about 1 in 10^7 flying hours tended to be viewed with concern by pilots and crews, although the overall accident rate, even under peace-time conditions, is much greater. For civil engineering structures the acceptable accident rate appears to be about 1 in 10^6 years, on which basis a building designed to have a life of 100 years should be allowed a probability of failure in its lifetime not exceeding about 1 in 10 000.

The other principal approach to the problem is on the grounds of economy. The overall cost may be expressed in the form

$$A + \psi_u \cdot C,$$

where

A = initial cost of structure,

C = capitalised cost of failure, i.e. the capital sum that must be invested at the outset to cover the cost of replacement of the structure,

ψ_u = probability of failure of structure during its lifetime.

For this is to be a minimum

$$\frac{\partial A}{\partial \psi_u} + \psi_u \frac{\partial C}{\partial \psi_u} + C = 0.$$

A and *C* will both be partly dependent on ψ_u, since a reduced probability of failure will require a stronger structure at an increased cost. If expressions can be set up for *A* and *C*, a mathematical solution of ψ_u is possible; a simple example of this is given by Pugsley [6.3] and a more detailed example of the design of reinforced concrete floor slabs for flats by Johnson [6.4]. More recently Sexsmith and Nelson [6.5] have shown how a minimum cost solution for ψ_u may be obtained by assessing the respective costs of failure and success for different levels of ψ_u using the 'decision tree' technique.

Evaluation of central load factor

It was shown above how the probability of failure could be calculated from the mean and standard deviation of the maximum load and of the ultimate resistance, respectively. The conventional central load factor may also be expressed in terms of these parameters by a formula due to Freudenthal [6.6] and its relationship to the probability of failure examined.

Let $\rho = D_m/s_D$ as defined above, i.e. the reciprocal of the coefficient of variation of $R - F$.

Since

$$\rho = \frac{R_m - F_m}{\sqrt{(s_R{}^2 + s_F{}^2)}},$$

$$\rho^2 s_R{}^2 + \rho^2 s_F{}^2 = R_m{}^2 - 2R_m F_m + F_m{}^2,$$

$$\frac{\rho^2 s_R{}^2}{R_m F_m} + \frac{\rho^2 s_F{}^2}{R_m F_m} = \frac{R_m}{F_m} - 2 + \frac{F_m}{R_m}.$$

Let γ_m be the central load factor relating the mean resistance and mean load, so that

$$\gamma_m = \frac{R_m}{F_m}.$$

Putting $1/F_m = \gamma_m/R_m$ in the above expression

$$\frac{\gamma_m \rho^2 s_R{}^2}{R_m} + \frac{\rho^2 s_F{}^2}{\gamma_m F_m{}^2} = \gamma_m - 2 + \frac{1}{\gamma_m},$$

$$\gamma_m{}^2 \left\{ 1 - \left(\frac{\rho s_R}{R_m}\right)^2 \right\} - 2\gamma_m + \left\{ 1 - \left(\frac{\rho s_F}{F_m}\right)^2 \right\} = 0,$$

and

$$\gamma_m = \frac{1 + \rho\sqrt{(\delta_R^2 + \delta_F^2 - \rho^2\delta_R\delta_F)}}{1 - \rho^2\delta_R^2},$$

where

$$\delta_R = \frac{s_R}{R_m} = \text{coefficient of variation of the ultimate resistance,}$$

$$\delta_F = \frac{s_F}{F_m} = \text{coefficient of variation of the maximum load.}$$

The central load factor is more commonly defined as the ratio of the characteristic ultimate resistance R_k to the characteristic maximum load F_k, where

$$F_k = F_m + \nu_{Fk}s_F = F_m (1 + \nu_{Fk}\delta_F),$$

$$R_k = R_m - \nu_{Rk}s_R = R_m (1 - \nu_{Rk}\delta_R).$$

The load factor γ_k is then given by

$$\gamma_k = \frac{R_k}{F_k} = \frac{(1 - \nu_{Rk}\delta_R)}{(1 - \nu_{Fk}\delta_F)}\gamma_m.$$

The above relationships are expressed graphically in Figs. 6.5 and 6.6 for the values $\nu_F = \nu_R = 1.64$, which are those normally adopted in the European Concrete Committee recommendations and the British Code.

It will be recalled from the previous section that the parameter ρ is related to ψ_u, the probability of failure, so that if the coefficients of variation of F and R can each be assigned a value, the value of the central load factor corresponding to a required probability of failure may be calculated or obtained from Figs. 6.5 and 6.6.

EXAMPLE

In a reinforced concrete column the coefficient of variation of the steel strength is 0.05 and that of the concrete strength 0.15. The coefficient of variation of the maximum load is 0.20. Calculate the central load factor for the probability of failure to be about 1 in 10^5.

The coefficient of variation of the ultimate resistance is given approximately by

$$\nu_R = 0.05^2 + 0.15^2 = 0.158,$$

$$\nu_F = 0.20.$$

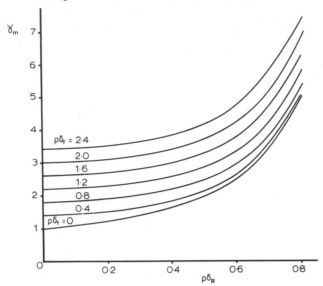

Figure 6.5   Relationship between coefficients of variation of load and resistance and central load factor

From tables of the normal distribution function, or from Fig. 6.4, for $\psi_u = 10^{-5}, \rho = 4.26$,

therefore

$$\rho v_R = 4.26 \times 0.158 = 0.675,$$

$$\rho v_F = 0.852.$$

v_m, the central load factor relating the mean maximum loads and the mean resistance, is obtained from Fig. 6.5 or from the formula

$$\gamma_m = \frac{1 + 4.26\sqrt{(0.158^2 + 0.20^2 - 0.675^2 \times 0.20^2)}}{1 - 0.675^2} = 3.57.$$

The allowed probability of the maximum load exceeding the characteristic value or of the ultimate resistance being less than the characteristic value is 5 per cent, so that $v_{Fk} = v_{Rk} = 1.64$.

The load factor relating the characteristic load and the characteristic ultimate resistance is therefore

$$\gamma_k = \left(\frac{1 - 1.64 \times 0.158}{1 + 1.64 \times 0.20}\right) 3.57$$

$$= 1.99$$

This may also be obtained from Fig. 6.6.

The ultimate resistance of many structural concrete members is mainly influenced either by the strength of the steel, as for example an under-reinforced beam, or by the strength of the concrete, as in an over reinforced beam or member subject to compression and bending. In such instances the coefficient of variation of the ultimate resistance of the member should be taken as that of the strength of the dominant material. Cornell [6.7] considers three coefficients of

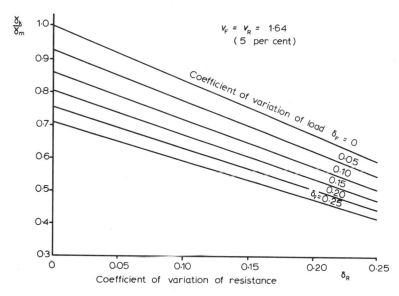

Figure 6.6 Relationship between central load factors for mean and characteristic values of load and resistance

variation, δ_m for the governing material strength, δ_f for fabrication (taking account of uncertainty in the construction process, e.g. dimensional inaccuracy), and δ_p for professional uncertainty in the assumptions adopted in the design calculations. The coefficient of variation of the ultimate resistance of the member is then given by

$$\delta_R = \sqrt{(\delta_m^2 + \delta_f^2 + \delta_p^2)}.$$

Evaluation of partial load factors

The single central load factor may be replaced by two or more partial factors in order to give the designer greater freedom to take account of special conditions. The simplest example is the use of two factors, one of which (γ_F) pertains to the load and the other (γ_R) to the resistance. The limiting condition relating the mean load and resistance is then

$$\gamma_F F_m = \frac{R_m}{\gamma_R} .$$

The central load factor is therefore

$$\gamma_m = \frac{R_m}{F_m} = \gamma_F \gamma_R .$$

Ravindra, Heaney and Lind [6.8] have presented approximate expressions for partial load factors in terms of the corresponding coefficients of variation, by the use of the linear approximation

$$(x^2 + y^2)^{\frac{1}{2}} \simeq \alpha(x + y)$$

where $0.707 < \alpha < 1$ and the error introduced by assuming $\alpha = 0.75$ is less than 10%

$$\rho = \frac{R_m - F_m}{\sqrt{(s_R^2 + s_F^2)}} = \frac{\gamma_m - 1}{\sqrt{(\gamma_m^2 \delta_R^2 + \delta_F^2)}}$$

$$\simeq \frac{\gamma_m - 1}{\alpha(\gamma_m \delta_R + \delta_F)} .$$

Hence

$$\gamma_m \simeq \frac{1 + \alpha\rho\delta_F}{1 - \alpha\rho\delta_R} = \gamma_F \gamma_R ,$$

where

$$\gamma_F = 1 + \alpha\rho\delta_F \quad \text{and} \quad \gamma_R = \frac{1}{1 - \alpha\rho\delta_R} .$$

When several variables contribute to the load or resistance the factors γ_F and γ_R may themselves be subdivided

$$\gamma_F = \gamma_{F1} \, \gamma_{F2} \, \gamma_{F3} \ldots \text{etc.}$$

$$\gamma_R = \gamma_{R1} \, \gamma_{R2} \, \gamma_{R3} \ldots \text{etc.}$$

For example, if $\gamma_R = \gamma_{R1} \, \gamma_{R2}$ and the corresponding coefficients of variation δ_{R1} and δ_{R2} so that $\delta_R = \sqrt{(\delta_{R1}^2 + \delta_{R2}^2)}$,

$$\gamma_R = \frac{1}{1 - \alpha\rho\sqrt{(\delta_{R1}^2 + \delta_{R2}^2)}}$$

$$\simeq \frac{1}{1 - \alpha\alpha_1\rho\delta_{R1} - \alpha\alpha_1\rho\delta_{R2}}$$

(using the above linear approximation)

$$= \frac{1}{(1 - \alpha\,\alpha_1\,\rho\,\delta_{R1})\,\left(1 - \dfrac{\alpha\,\alpha_1\,\rho\,\delta_{R2}}{1 - \alpha\,\alpha_1\,\delta_{R2}}\right)}$$

$$= \gamma_{R1}\,\gamma_{R2},$$

where

$$\gamma_{R1} = \frac{1}{1 - \alpha_{R1}\,\rho\,\delta_{R1}} \quad \text{and} \quad \gamma_{R2} = \frac{1}{1 - \alpha_{R2}\,\rho\,\delta_{R2}}$$

in which α_{R1} and α_{R2} are sensibly constant.

Limit state design

Failure has so far been discussed in terms of total collapse, considering the probability of the applied load exceeding the ultimate resistance of a member. It has already been noted, however, that serviceability, no less than safety, is a requirement in structural design, and that permissible stresses have sometimes been specified from this consideration. A more rational approach is to base the design on the resistance of the member at the point at which it may be deemed to become unserviceable. This may occur in a number of possible ways; for example, the deflexion may exceed the acceptable value, the width of flexural cracks may become excessive, or there may be local spalling of the concrete. If the resistance corresponding to a particular mode of unserviceability can be calculated, one can treat serviceability in terms of the risk of the applied load exceeding this resistance, either by conventional load factors or by introducing the more sophisticated probability procedures already outlined. The methods for safety and serviceability will thus become similar and the ultimate resistance will be one of a number of possible limit states. A higher load factor (or lower probability) will of course be specified against the load exceeding the ultimate limit state than will be specified against the load exceeding one of the serviceability limit states.

Criteria for limit states

Limit state design, as defined above, has been adopted for the British Code, following the recommendations of the European Concrete Committee. Two principal serviceability limit states are defined, namely deflexion and cracking.

In certain special circumstances other limit states may be important, such as vibration under dynamic loading.

Each limit state may be attained in a number of possible ways, although in practice only one or two of these will be significant and it will usually be clear which can be ignored in a particular instance. Some of the criteria for the ultimate limit state are as follows:

Bearing failure at a support or under a concentrated load
Bond or anchorage failure of reinforcement
Bursting of prestressed concrete end blocks
Elastic instability of a member
Failure of connexions (e.g. between pre-cast concrete elements or in composite construction)
Flexural failure at one or more critical sections
General instability (e.g. overturning)
Shear failure
Torsion failure

Serviceability limit states - deflexion

It is difficult to establish a precise value of maximum permissible deflexion on rational grounds and any proposals are inevitably rather arbitrary. The deflexion may have to be limited on account of the discomfort experienced by users of a structure which is very flexible, but before this point is reached damage is likely to be caused to building finishes such as ceilings and partitions. Extreme instances of lateral deflexion in columns may affect the ultimate limit state.

The following are the recommendations of the British Code:

(a) Deflexion of Beams not to exceed Span/250 including the effect of temperature shrinkage and creep.
(b) Additional deflexion after completion finishes and partitions not to exceed Span/350 or 20 mm (0.8 in), whichever is less.
(c) The preformed upward camber of beams not to exceed Span/300.

Cracking and spalling

Local damage to structural concrete members may take the form of cracks of such widths as to be unsightly or to expose the reinforcement to corrosion; or there may be local crushing or spalling of the concrete.

According to the British Code the width of crack, measured at the surface of a member, should not exceed 0.3 mm (0.012 in), or 0.004 times the minimum cover at the main reinforcement where there is an aggressive environment. The security against cracking, or limitation of the crack width in prestressed members is governed by restriction of the tensile stress in the concrete according to three classes of structure as explained in Chapter 2 (p. 65f) and Chapter 7 (p. 190 f):

Class 1 — no tensile stress permitted,
Class 2 — tensile stress permitted but no visible cracking,
Class 3 — cracking permitted, but width not to exceed 0.2 mm (0.008 in) or 0.1 mm (0.004 in) in an aggressive environment. This class has been variously termed partial prestressing, limited prestressing, or prestressed reinforced concrete.

Limit state design by partial factors

The practical application of probability methods is at present only feasible in a small number of instances. Sexsmith and Nelson [6.5] have summarised the difficulties, chief among which is the absence of the necessary statistical data, particularly of the frequency of load occurrence. A further difficulty arises from the uncertainty involved in applying the results of structural analysis and resistance calculations, or of laboratory tests, to full scale structures. Thirdly the probability theory is itself open to considerable doubt, as the calculation of the probability of failure depends on a few extreme values of load and resistance at the 'tail' of the distribution curve. In limit state design, there is the added problem of defining other limit states than that of collapse, and calculating the appropriate resistance.

Having regard to these limitations the European Concrete Committee have proposed a method of limit state design by the use of partial factors, which has been adopted in the British Code. The different factors can be varied to take account of different loading and structural conditions and may be 'recalibrated' by probability theory as and when the necessary statistical information becomes available.

The first set of partial factors $(\gamma_{f1}, \gamma_{f2}, \ldots \gamma_{fn})$ are applied as multipliers to the components $(F_{k1}, F_{k2}, \ldots F_{kn})$ of the characteristic load (e.g. dead load, imposed load, wind load, etc.) as in the load factor method of design described earlier. The design load for a particular limit state is therefore $\gamma_{f1}F_{k1} + \gamma_{f2}F_{k2} + \gamma_{fn}F_{kn}$ and the design resistance must be greater than this load. The design resistance is calculated by substituting in the appropriate resistance formula factored values of the resistance of the consituent materials, obtained by dividing the characteristic resistance of each material, R_{k1}, R_{k2},

$\ldots$, R_{kn} by an appropriate 'materials factor', γ_{m1}, γ_{m2}, $\ldots$ γ_{mn}. Thus if the resistance of the member is a function of the resistances of the constituent materials, $f(R_1, R_2, \ldots R_n)$ the design resistance will be

$$ f\left(\frac{R_{k1}}{\gamma_{m1}}, \quad \frac{R_{k2}}{\gamma_{m2}}, \quad \ldots \quad \frac{R_{kn}}{\gamma_{mn}}\right). $$

The basic design condition is therefore

$$ \gamma_{f1}F_{k1} + \gamma_{f2}F_{k2} + \quad \ldots \quad \gamma_{fn}F_{kn} \leqslant f\left(\frac{R_{k1}}{\gamma_{m1}}, \quad \frac{R_{k2}}{\gamma_{m2}}, \quad \ldots \quad \frac{R_{kq}}{\gamma_{mq}}\right). $$

It has already been shown how the characteristic resistance of a material is statistically determined and the characteristic loads may be similarly defined. Further information may in future enable the partial factors γ_F and γ_m to be adjusted, having regard to the overall probability of failure; at present, however, they have been established to give results largely similar to those obtained with the existing codes.

Values of partial factors

The design values of the partial load factors γ_f recommended in the British Code are given in Table 6.1 for the two principal limit states. It will be seen that the factor is generally unity for the serviceability limit state. This may appear illogical, suggesting as it does that there is no margin of security against these limit states. However, the way in which the characteristic loads and resistances are defined does in fact provide such a margin, and the factor of unity provides a welcome simplification.

The components of the load have to be considered in their most unfavourable combinations, in which the imposed load is not always a maximum but may be zero as on alternate spans of a continuous beam, and for the same reason the dead load factor is reduced when the dead load is in combination with wind load.

The materials factor γ_m generally recommended for concrete at the ultimate limit state is 1.5, but according to the European Concrete Committee's recommendations it could vary between 1.4 and 1.6 depending on the standard of supervision. This refers to placing, compaction and curing on site but the standard of materials, batching and mixing are determined when establishing the characteristic strength. For the limit states of deflexion and cracking the respective factors are 1.0 and 1.3.

Table 6.1   *Design Values of Load Factor γ_f (British Code)*

Limit state	Dead load			Imposed load		Wind load	
	With imposed load only	With wind load only	With imposed and wind load	With dead load only	With dead and wind load	With dead load only	With dead and imposed load
Ultimate	1.4	0.9	1.2	1.6	1.2	1.4	1.2
Serviceability	1.0	1.0	0.8	1.0	0.8	1.0	0.8

Steel being generally less liable to variation than concrete, the factor is taken as 1.15 for the ultimate limit state and unity for the other limit states. In prestressed members the prestressing force may either be increased or decreased by errors of construction and a factor of 1.1 is recommended by the European Concrete Committee for the majority of instances.

The formulae in the British Code represent the design resistance for various types of member. These are written in terms of the characteristic resistance values for the materials and the materials factors γ_m have been incorporated in the numerical coefficients. A number of examples of these factored design resistance formulae have already been given in Chapters 4 and 5.

REFERENCES

[6.1]   Corso, J. M. (1954) Discussion to reference 6.6 *Proc. Amer. Soc. Civ. Engrs,* **80,**

[6.2]   Su, H. L. (July 1959) Statistical approach to structural design. *Proc. Instn Civ. Engrs.,* **13,** 353-362.

[6.3]   Pugsley, A. G. (1966) *The Safety of Structures,* Ch. 9. London: Arnold.

[6.4]   Johnson, A. I. (1956) The determination of the design factor for reinforced concrete structures. In *Symposium on the Strength of Concrete Structures,* London: Cement and Concrete Assocn.

[6.5]   Sexsmith, R. G. and Nelson, M. F. (Oct. 1969) Limitations in application of probabilistic concepts. *Proc. Amer. Concr. Inst.,* **66,** 823-828.

[6.6]   Freudenthal, A. M. (1956) Safety and the probability of structural failure. *Trans. Amer. Soc. Civil Engnrs,* **121,** 1337-1397.

[6.7]   Cornell, C. A. (Dec. 1969) A probability-based structural code. *Proc. Amer. Concr. Inst.,* **66,** 974-985.

[6.8]   Ravindra, M. K., Heaney, A. C., and Lind, N. C. (1969) Probabilistic evaluation of safety factors. In *Symposium on concepts of safety of structures and methods of design.* London: Int. Assoc. Bridge & Struct. Engng.

[6.9]   Entroy, H. C. (1960) The variation of works test cubes. *Research Report No. 10.* London: Cement and Concrete Association.

7

Dimensioning for simple bending

Procedure for dimensioning

The dimensioning of structural concrete members will now be considered on the basis of the limit state principles described in the previous chapter, using the method of partial factors.

In Part I the calculations were of an analytical nature, commencing from the known dimensions of the member and resistance properties of the constituent materials, steel and concrete. From these data one proceeded to determine the resistance of the member corresponding to a particular state, e.g. failure in a given mode, or a given degree of crack width or deflexion. However, in the design process now under discussion, the minimum required resistance is pre-determined for certain limit states, such as ultimate failure, deflexion and cracking, for each of which the design value of the resistance must equal or exceed the sum of the specified characteristic loads each multiplied by the appropriate partial load factor. The resistance of the materials is also specified, and is to some extent under the control of the designer, the values used being the characteristic resistance of each material divided by the appropriate partial factor. The problem is then to determine suitable dimensions for the member — the cross section, area of reinforcement, amount of prestress, etc. — such that the design resistance is attainable for the limit state in question.

Obviously one possible method is by using an analytical formula to check the resistance of a member of assumed dimensions, correcting the latter until a satisfactory result is obtained. This 'trial and error' procedure is quite often adopted in practice, and is sometimes essential. It is, however, often wasteful of time and a direct method of calculating the required dimensions is usually to be preferred.

The need to satisfy three limit state conditions might appear to be a complication, but, as will be shown, each limit state often governs a different set of dimensions of the same member, and the complete dimensions of the member may be built up by considering the limit states in turn. In other instances a member may be directly designed for one limit state and checked for the others.

Leading dimensions of a section — ultimate limit state

In both reinforced and prestressed concrete members the necessary effective depth (d) and breadth of a section at the compression face (b) are controlled mainly by the ultimate limit state. If the section is of I or T shape the required depth of the top flange or slab (h_f) is also largely dependent on this limit state.

It was shown in Chapter 4 (p. 102) that according to the British Code the maximum design value of the moment of resistance of a rectangular reinforced concrete section without compressive reinforcement was given by

$$M_{ud} = 0.15 f_{cu} bd^2 .$$

This corresponds to an assumed maximum depth of the compressive zone $x = 0.5d$. Similar expressions may be derived for prestressed sections. The maximum value of x/d for which provision is made in the code is 0.783 with pre-tensioned tendons and 0.653 with post-tensioned tendons (Fig. 4.13). Correspondingly the maximum design moment of resistance with pre-tensioned tendons is

$$M_{ud} = 0.6 \left(\frac{f_{cu}}{1.5}\right) \times 0.783 \, (1 - 0.5 \times 0.783) \, bd^2$$

$$= 0.190 \, f_{cu} \, bd_1{}^2 \, \phi,$$

and with post-tensioned tendons

$$M_{ud} = 0.175 \, f_{cu} bd^2 .$$

Since the above expressions may be satisfied by a whole range of possible combinations of b and d, the designer has some freedom of choice in deciding these values. In some instances d may be restricted by a limited construction depth, while on other occasions b may be determined by practical considerations. Otherwise values of the two dimensions may be selected so as to give reasonable proportions.

It must be appreciated that the dimensions determined by the above expressions are minimum values and it will often be preferable to use a larger section with a resultant saving of reinforcement or prestressing tendons. It is possible to use a smaller section by the introduction of compressive reinforcement, but if the depth is small, care must be taken that the limit state of deflexion is not exceeded.

The maximum design moment of resistance of an I or T section will usually be less than that of a rectangular section having the same effective depth and breadth at the compression face. The British Code gives a conservative formula for reinforced concrete T sections, referred to in Chapter 4 (p. 107), neglecting the resistance of the concrete below the top slab, namely

$$M_{ud} = 0.4 f_{cu} \, b \, h_f \left(d - \frac{h_f}{2} \right)$$

The more general formula, applicable to prestressed T and I sections is

$$M_{ud} = f_{cu} b d^2 \left[\frac{b_w}{b} \cdot \frac{x}{d} \left(0.4 - 0.2 \frac{x}{d} \right) + \left(1 - \frac{b_w}{b} \right) \frac{h_f}{d} \left(0.4 - 0.2 \frac{h_f}{d} \right) \right].$$

The maximum moment is obtained by inserting the appropriate maximum values of x/d given above. With $(x/d) = 0.5$ the formula becomes

$$M_{ud} = f_{cu} b d^2 \left[0.15 \frac{b_w}{b} + \left(1 - \frac{b_w}{b} \right) \frac{h_f}{d} \left(0.4 - 0.2 \frac{h_f}{d} \right) \right].$$

In Fig. 7.1 the above relationship is shown graphically for a range of values of h_f/d and b_w/b. This type of chart is a useful guide in the preliminary dimensioning of both reinforced and prestressed concrete sections.

Allowance for self-weight

It is necessary to include an estimated self-weight of the member in the total load when calculating the dimensions of the section. The self-weight constitutes a relatively greater proportion of the load as the span increases, and the estimating of its value may present a problem until some experience has been obtained. Inaccuracy in this estimate will necessitate a revision of the preliminary calculations.

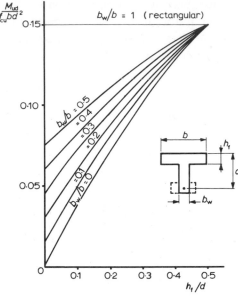

Figure 7.1   Chart for preliminary dimensioning of reinforced
and prestressed concrete sections

An approximate guide to the self-weight of a member of rectangular section
may be obtained as follows:

$$M_{ud} = 0.15\, f_{cu}bd^2 < \beta\, w_{ud}\, l^2$$

where w_{ud} is the total uniformly distributed load corresponding to the design
moment of resistance M_{ud} for the ultimate limit state, and β is the moment
coefficient (e.g. 0.125 for a simply supported beam).

The self-weight w_{min}, sustained throughout the life of the member is given,
for a rectangular section, by $w_{min} = bhD_c g$ where D_c is the density of the
member and g the gravitational acceleration which must be included in the S.I.
system but is omitted when using gravitational units of force (e.g. lbf). Hence

$$\frac{w_{min}}{w_{ud}} = \frac{D_c g\, \beta(l/h)\, l}{0.15\, f_{cu}\, (d/h)^2}.$$

Increasing the numerical constant to cover errors due to the rounding off of
dimensions

$$\frac{w_{min}}{w_{ud}} = \frac{7.5\, D_c g\, \beta(l/h)\, l}{f_{cu}\, (d/h)^2}$$

In using this formula l/h, the ratio span/overall depth, may be given a typical
maximum value, say 20 for a simply supported beam, 30 for a fixed-ended beam

and 10 for a cantilever. The ratio d/h usually lies between 0.85 and 0.95 and where there is uncertainty the lower value should be used. For flanged T or I sections w_{min} is about one half of the value given by the above formula, and a numerical constant of 4 is suitable.

The design load w_{ud} at the ultimate limit state includes the self-weight enhanced by a factor γ_g. Thus the required design ultimate load may be calculated by dividing the sum of the factored loads (excluding the self-weight) by $1 - \gamma_g (w_{min}/w_{ud})$, having estimated w_{min}/w_{ud} from the above formula.

EXAMPLE

The above and subsequent steps in dimensioning will be illustrated by working an example of a simply-supported beam, first in reinforced concrete with a rectangular section and then in reinforced concrete with a T section and in prestressed concrete using an unsymmetrical flanged section.

The data are as follows:

Span: 25 m (82 ft).

Breadth of top flange: 600 mm (23½ in).

Dead load (in addition to self weight): 2 kN/m (135 lbf/ft).

Imposed live load: 15 kN/m (1030 lbf/ft).

Characteristic strength of concrete: 45 N/mm² (6 500 lbf/in²) at 28 days and 37.5 N/mm² (5 400 lbf/in²) at transfer of prestress.

Characteristic yield stress of reinforcement: 420 N/mm² (61 000 lbf/in²).

Characteristic tensile strength of tendons: 1500 N/mm² (218 000 lbf/in²).

Loss of prestress after transfer: 15%.

(a) Reinforced concrete beam of rectangular section

$$w_{min}/w_{ud} = \frac{7.5 \times 2\,400 \times 9.81 \times 0.125 \times 20 \times 25}{45 \times 10^6 \times 0.85^2} = 0.34.$$

Ultimate load excluding factored self-weight

$$= 1.4 \times 2 + 1.6 \times 15 = 26.8 \text{ kN/m (1830 lbf/ft)}.$$

$$w_{ud\ req.} = \frac{26.8}{1 - 1.4 \times 0.34} = 51.2 \text{ kN/m (3 500 lbf/ft)}$$

$$M_{ud\ req.} = 51.2 \times 25^2 \times 0.125 = 4\,000 \text{ kN. m}$$

(35 300 000 kNm)

$$0.15 \times 45\ bd^2 \geqslant 4\,000 \times 10^6,$$

therefore

$$bd^2 \geqslant 592 \times 10^6 \text{ mm}^3 \ (36\ 200 \text{ in}^3).$$

Since b is to be 600 mm the minimum possible value of d would be about 1000 mm so that $bd^2 = 600 \times 10^6$ (36 600 in^3). The overall depth would then be about $1000/0.85 = 1175$ mm (46.2 in) which might be rounded off to 1200 mm (48 in) (Fig. 7.2(a)). The actual self-weight would then be

$$w_{min} = 0.6 \times 1.2 \times 2400 \times 9.81 \times 10^{-3} = 17.0 \text{ kN/m (1160 lbf/ft)}.$$

This may be compared with the allowance made in the calculation, where

$$w_{min} = 0.34 \times 51.2 = 17.4 \text{ kN/m (1190 lbf/ft)}.$$

The span/depth ratio is $25/1.2 = 20.8$, slightly above the initially assumed value of 20, but the limit state of deflexion will later be checked.

(b) Reinforced concrete T beam

The above solution represents the minimum depth without compressive reinforcement. If conditions allow a greater depth of beam a deep T section may be adopted with a resultant saving of self-weight and reinforcement. Here the overall depth will be increased to 1400 mm (55 in) and the effective depth assumed to be $0.85 \times 1400 = 1190$ mm (47 in)

$$l/h = 25/1.4 = 17.8,$$

$$w_{min}/w_{ud} = \frac{6 \times 2400 \times 9.81 \times 0.125 \times 17.8 \times 25}{45 \times 10^6 \times 0.85^2} = 0.24,$$

$$w_{ud\ req.} = \frac{26.8}{1 - 1.4 \times 0.24} = 40.4 \text{ kN/m (2760 lbf/ft)},$$

$$M_{ud\ req.}\ 40.4 \times 25^2 \times 0.125 = 3160 \text{ kN m (28 000 000 lbf in)}.$$

$$\frac{M_{ud\ req.}}{f_{cu}\ b\ d^2} = \frac{3160 \times 10^6}{45 \times 600 \times 1190^2} = 0.083.$$

Using the appropriate formula from the British Code,

$$3160 \times 10^6 = 45 \times 600 \times 1190^2 \left\{ 0.4 \left(\frac{h_f}{d} \right) - 0.2 \left(\frac{h_f}{d} \right)^2 \right\}.$$

Solving the equation, or more rapidly by using Fig. 7.1 with $b_w/b = 0$, $h_f/d = 0.23$, so that $h_f = 0.23 \times 1190 = 274$ mm (10.8 in).

A lighter member can be designed, by taking into account the contribution of the upper part of the rib to the compressive resistance of the concrete. The width of the rib b_w may be determined by the ultimate limit state in shear to be discussed later, and must also be sufficient for the reinforcement, which includes bent-up bars when applicable. If a width of 250 mm is selected so that $b_w/b = 250/600 = 0.41$, h_f/d is found from Fig. 7.1 to be about 0.10. Thus a suitable value of h_f would be 1190 x 0.10, i.e. approximately 120 mm (43/4 in).

The section of the beam is shown in Fig. 7.2(b). The self-weight is 9.2 kN/m (630 lbf/ft) compared with a load of 9.7 kN/m allowed in the calculation.

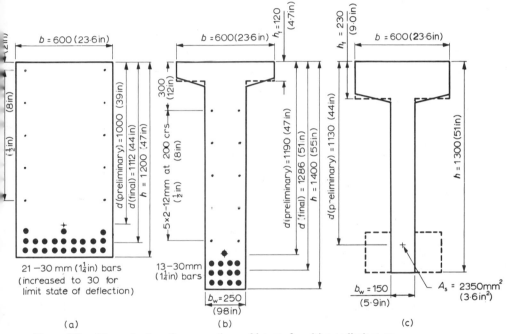

Figure 7.2 Dimensioning of cross sections of beams for ultimate limit state

(c) Prestressed concrete beam

Since the deflexion of an uncracked prestressed beam is less than that of a reinforced concrete beam, the estimated l/h ratio is increased to 25.

$$\frac{w_{min}}{w_{ud}} = \frac{4 \times 2\,400 \times 9.81 \times 0.125 \times 25 \times 25}{45 \times 10^6 \times 0.85^2} = 0.23.$$

$$w_{ud\ req.} = \frac{26.8}{1 - 1.4 \times 0.23} = 39.6 \text{ kN/m (2700 lbf/ft)}.$$

$$M_{ud\ req.} = 39.6 \times 25^2 \times 0.125 = 3\,100 \text{ kn m (27 400 000 lbf in)}$$

The value of $M_{ud\ req.}/bd^2$ for a flanged section is less than 0.15 and may be estimated from Fig. 7.1. A reasonable average value for h_f/d is about 0.2 while b_w the breadth of the web is largely determined by the space required to contain the ducts for the tendons with adequate side cover. It would be about 150 mm (6 in) in a beam of this size so as to accommodate a 50 mm (2 in) duct with 50 mm (2 in) for web reinforcement and side cover. Thus $b_w/b = 0.25$ and from Fig. 7.1

$$\frac{M_{ud}}{f_{cu}\ b\ d^2} \simeq 0.09.$$

Thus

$$d > \frac{3100}{45 \times 600 \times 0.09} = 1130 \text{ mm (44.5 in)},$$

$$h \simeq \frac{1130}{0.85} = 1340 \text{ mm, say } 1300 \text{ mm (51 in)},$$

$$h_f = 1130 \times 0.2 = 226 \text{ mm, say } 230 \text{ mm (9 in)}.$$

The section dimensioned thus far is shown in comparison with the two reinforced concrete sections in Fig. 7.2 (c).

Total tensile reinforcement

The next step is to calculate the total area of tensile reinforcement, which is also governed by the ultimate limit state. It is preferable to use a general expression applicable to any size of section rather than to consider only a section of minimum dimensions. The depth of the compression zone x must therefore be obtained.

Re-writing equation 4.1.1(b)′, Chapter 4 (p. 96) and inserting the factors and coefficients assumed in the British code

$$\frac{M_{ud}}{f_{cu}\ b\ d^2} = 0.4\ \xi\ (1 - 0.5\ \xi).$$

$$\frac{x}{d} = \xi = 1 - \sqrt{(1 - 5M_{ud}/f_{cu}bd^2)}$$

Then from equation 4.1.1(a)

$$\frac{A_s}{bd} = \frac{0.4\ f_{cu}\xi}{0.87\ f_{sb}} = \frac{0.46\ f_{cu}}{f_{sb}} \left\{ 1 - \sqrt{\left(1 - \frac{5M_{ud}}{f_{cu}\ b\ d^2}\right)} \right\},$$

or alternatively

$$A_s = \frac{M_{ud}}{0.87\, f_{sb}(d - 0.5x)}\,.$$

f_{sb} is the stress developed in the reinforcement, assumed in reinforced concrete to be equal to f_y, the yield stress or its equivalent (e.g. 0.2% proof stress). In prestressed concrete the ratio of this stress to the tensile strength f_{pu} of the steel tendons is considered to be related to x/d in accordance with the empirical relationship shown in Fig. 4.13.

A_s represents the total area of tensile reinforcement required. In a prestressed member this is provided mainly by the tendons, but some supplementary reinforcement may also be necessary to satisfy the ultimate limit state condition. A_s is therefore replaced by A_{se}, the effective area of steel of strength f_{pu} used for the tendons, which may be considered to be given by

$$A_{se} = A_{ps} + \frac{f_y}{f_{pu}} A_s$$

where A_s is the area and f_y the yield stress of the supplementary untensioned reinforcement. The area of the tendons A_{ps} will later be calculated for the serviceability limit state, enabling A_s to be determined.

The area of tensile reinforcement in a reinforced concrete T section may be calculated by the approximate formula

$$A_s = \frac{M_{ud}}{0.87\, f_y\,(d - h_f/2)}\,.$$

This assumes that the centre of compression is located at the mid-depth of the top slab or flange. Where it is also necessary to consider the rib, M_{ud} is divided into M_{fud} the component provided by the projecting part of the top flange (i.e. a breadth $b - b_w$), and into $M_{ud} - M_{fud}$, the component provided by a rectangular section of breadth b_w.

Then

$$M_{fud} = 0.4\, f_{cu}\,(b - b_w)\, h_f \left(d - \frac{h_f}{2}\right),$$

$$\frac{x}{d} = 1 - \sqrt{\left(1 - \frac{5(M_{ud} - M_{fud})}{f_{cu} b_w d^2}\right)}.$$

The total area of steel is the sum of the areas required for M_{fud} and $M_{ud} - M_{fud}$ respectively.

$$A_s = \frac{0.46 f_{cu} (b - b_w) h_f}{f_{sb}} + \frac{0.46 f_{cu} b_w x}{f_{sb}},$$

or alternatively

$$A_s = \frac{M_{fud}}{0.87 f_{sb}(d - h_f/2)} + \frac{M_{ud} - M_{fud}}{0.87 f_{sb}(d - x/2)} .$$

For a prestressed member f_{sb} is obtained from x/d as above.

EXAMPLE

Continuing the previous examples,

(a) Reinforced concrete beam of rectangular section

In this example, in which the effective depth is close to the minimum value, the depth of the compressive zone will be approximately equal to the maximum

$$x = \frac{d}{2} = \frac{1000}{2} = 500 \text{ mm (19.7 in)}.$$

The area of tensile reinforcement is given approximately by

$$A_s = 0.23 \frac{f_{cu} bd}{f_y} = \frac{0.23 \times 45 \times 600 \times 1000}{420} = 14\ 800 \text{ mm}^2 \ (23.0 \text{ in}^2).$$

The exact calculation is as follows:

$$\frac{x}{d} = 1 - \frac{5 \times 4000 \times 10^6}{45 \times 600 \times 1000^2} = 0.492,$$

$$A_s = \frac{4000 \times 10^6}{0.87 \times 420 \times 1000} (1 - 0.492/2) = 14\ 500 \text{ mm}^2 \ (22.5 \text{ in}^2).$$

21-30 mm bars would give an area of 14 800 mm² (19-1¼ in bars would give 23.3 in²) and may be arranged as shown in Fig. 7.2(a), so that the actual effective depth is 1112 mm.

The distribution of the tensile reinforcing bars is important if the crack width is not to exceed the value associated with the serviceability limit state, since cracking is governed mainly by the strain in the reinforcement and the cover and spacing of the bars. The former is restricted by the stress allowed in the steel at the ultimate limit state while the British Code recommends that the clear distance between bars should not exceed 150 mm (6 in), and the cover measured to a corner should not exceed 70 mm (2¾ in). A further requirement is that where the depth of the beam exceeds 750 mm (30 in) there should be longitudinal bars at the sides of the beam, the clear distance between them not exceeding 200 mm (8 in). In the present example it is therefore necessary to place additional reinforcement at the sides of the beam for crack control, for which 12 mm (½ in) bars at about 100 mm centres would be suitable. Where the crack width is likely to be critical it may be checked by the formula given in Chapter 2 (p. 63f).

(b) Reinforced concrete T beam

The area of reinforcement for the section designed by the approximate formula is

$$A_s = \frac{3160 \times 10^6}{0.87 \times 420\,(1190 - 274/2)} = 8200 \text{ mm}^2 \ (12.7 \text{ in}^2).$$

For the section designed by taking the web into account,

$$M_{fud} = 0.4 \times 45\,(600 - 250)\,120\,(1190 - 250/2) \times 10^{-6}$$

$$= 805 \text{ kN m } (7\,120\,000 \text{ lbf in}).$$

$$\frac{x}{d} = 1 - \sqrt{\left\{1 - \frac{5(3160 - 805) \times 10^6}{45 \times 250 \times 1190^2}\right\}} = 0.49,$$

$$x = 0.49 \times 1190 = 584 \text{ mm } (23.0 \text{ in}),$$

$$A_s = \frac{805 \times 10^6}{0.87 \times 420\,(1190 - 120/2)} + \frac{(3160 - 805) \times 10^6}{0.87 \times 420\,(1190 - 584/2)}$$

$$= 9140 \text{ mm}^2 \ (14.2 \text{ in}^2).$$

13-30 mm bars give 9150 mm² (12-1¼ in bars give 14.8 in²) and may be arranged as in Fig. 7.2b.

(c) Prestressed concrete beam

$$M_{fud} = 0.4 \times 45\,(600 - 150) \times 230\,(1130 - 230/2) \times 10^6 = 1870 \text{ kN m}$$
$$(16\,500\,000 \text{ lbf in}).$$

$$\frac{x}{d} = 1 - \sqrt{\left\{1 - \frac{5\,(3100 - 1870) \times 10^6}{45 \times 150 \times 1130^2}\right\}} = 0.46.$$

$$x = 0.46 \times 1130 = 520 \text{ mm } (20.4 \text{ in}).$$

$$A_{se} = \frac{1870 \times 10^6}{0.87 \times 1600\,(1130 - 230/2)}$$

$$+ \frac{(3100 - 1870) \times 10^6}{0.87 \times 1600\,(1130 - 520/2)} = 1335 + 1015 = 2350 \text{ mm}^2 \ (3.64 \text{ in}^2).$$

This is an equivalent area of high strength steel, part of which will be provided by the tendons and the remainder by supplementary reinforcement.

Dimensions controlled by deflexion

In certain instances the leading dimensions of a section may have to be chosen to satisfy the conditions for the limit state of deflexion. Where this is likely to be important it is suggested that a preliminary calculation be made of the second moment of area of the concrete section I_c required to satisfy the deflexion equation.

$$a = \frac{\beta l^2 M}{E_c I_c},$$

where M is the moment and a the deflexion at the limit state.

For a reinforced concrete section with tensile reinforcement,

$$I_c = \alpha \rho (1 - \xi)(1 - \xi/3)bd^2.$$

The curves for this expression are given for a series of values of the modular ratio in Fig. 7.3 and can be used to give an indication of the size of section and area of reinforcement required. The value of I_c will be somewhat greater for a section with compressive reinforcement and less for a T section.

A prestressed member will normally be uncracked at the load corresponding to the limit state of deflexion, so that I_c will be based on the full cross section. The exact value will, of course, depend on the dimensions of the flanges and web, as well as on the area of the tendons and reinforcement, but an estimate of the required overall dimensions b and h may be obtained from the following approximations:

for a rectangular section

$$I_c \simeq bd^3/12,$$

for an I section (symmetrical)

$$I_c \simeq 0.7\,(bd^3/12),$$

for a T section (without bottom flange)

$$I_c \simeq 0.4\,(bd^3/12).$$

EXAMPLE

Choose the dimensions of the three beams in the previous example so that the final deflexion is not likely to exceed 1/125 of the span (1/250 pre-camber and 1/250 below support level).

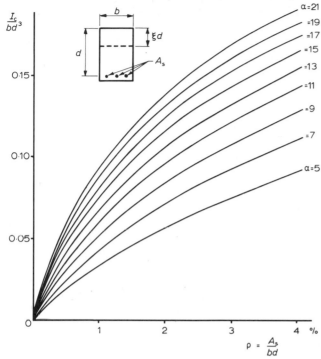

Figure 7.3   Second moment of area of cracked concrete sections with reinforcement

The limiting deflexion is 25 × 1000/125 = 200 mm. According to the British Code the short-term modulus of elasticity of concrete having a characteristic compressive strength of 45 kN/mm², is assumed to be 32.5 N/mm² (4.35 × 10⁶ lbf/in²).

Modular ratio = 200/30 = 6.7.

(a) Reinforced concrete beam of rectangular section
Permanent (dead) load at limit state of deflexion

$$w_g = 17.0 \times 1.0 + 2.0 \times 1.0 = 19.0 \text{ kN/m (1300 lbf/ft).}$$

Total design load

$$w_d = 17.0 \times 1.0 + 2.0 \times 1.0 + 15.0 \times 1.0 = 34.0 \text{ kN/m (2320 lbf/ft).}$$

$$M_d = 34.0 \times 25^2 \times 0.125 = 2660 \text{ kN m (23 500 000 lbf in).}$$

The additional deflexion of a beam without compressive reinforcement may be assumed to be 1.8 times the instantaneous deflexion due to the permanent load, when loading commences at 28 days. Therefore, if a is the total instantaneous deflexion

$$200 = a + \frac{19.0}{34.0} a \times 1.8$$

$$a = 100 \text{ mm } (3.9 \text{ in}).$$

The deflexion equation is therefore

$$100 > \frac{(5/48) \times 25^2 \times 10^6 \times 2660 \times 10^3}{32.5 \, I_c},$$

therefore

$$I_c > 53\ 300 \times 10^6 \text{ mm}^4 \ (128\ 500 \text{ in}^4)$$

$$\frac{I_c}{bd^3} > \frac{53\ 300 \times 10^6}{600 \times 1000^3} = 0.089.$$

From Fig. 7.3 it is seen that with $\alpha = 6.2$, I_c/bd^3 will be greater than 0.089 provided that the tensile steel ratio is greater than about 3.5%. Since the reinforcement ratio required for the ultimate limit state is about 2.4% the deflexion will be critical and the number of bars will need to be increased from 21 to 30. A more economical solution, however, would be to design a section of greater depth with a smaller amount of reinforcement.

(b) Reinforced concrete T beam

In this example I_c is required to be $36\ 300 \times 10^6$ mm^4 (87 200 in^4) so that $I_c/bd^3 = 0.036$, but the reinforcement ratio required will be greater than the value for a rectangular section given by Fig. 7.3. An estimate may be obtained by assuming an average breadth of 400 mm (16 in).

$$I_c/bd^3 = \frac{36\ 300 \times 10^6}{400 \times 1190^3} = 0.054.$$

From Fig. 7.3 $\rho \geqslant 1.6\%$,

$$A_s \geqslant \frac{1.6 \times 400 \times 1190}{100} = 7610 \text{ m}^2 \ (11.8 \text{ in}^2).$$

The 13-30 mm bars required for the ultimate limit state have an area of 9150 mm^2 and should therefore satisfy the limit state condition for deflexion.

(c) *Prestressed concrete beam*

The estimated self-weight (p. 177) is

$$w_{min} = 0.23 \times 39.6 = 9.1 \text{ kN/m (625 lbf/ft)}.$$

Permanent load at limit state of deflexion

$$w_g = 9.1 \times 1.0 + 2.0 \times 1.0 = 11.1 \text{ kN/m (760 lbf/ft)}.$$

Total design load

$$w_d = 9.1 \times 1.0 + 2.0 \times 1.0 + 15.0 \times 1.0 = 26.1 \text{ kN/m (1780 lbf/ft)}.$$

$$M_d = 26.1 \times 25^2 \times 0.125 = 2040 \text{ kN m (18 050 000 lbf in)}.$$

Assuming the time-dependent deflexion to be 1.8 times the instantaneous value as for reinforced concrete

$$a = 113 \text{ mm (4.4 in)}.$$

From the deflexion equation

$$I_c = 36\ 200 \times 10^6 \text{ mm}^4 \text{ (87 000 in}^4\text{)}.$$

$$\frac{I_c}{bd^3/12} = \frac{36\ 200}{600 \times 1300^3/12} = 0.33.$$

This ratio will be exceeded in a normal section. It should be noted that part of the pre-camber will be provided by the initial upward deflexion due to the prestress.

Compressive reinforcement

The moment of resistance of a member may be increased beyond the value corresponding to the maximum compressive resistance of the concrete by the use of compressive reinforcement. The area of compressive reinforcement required is given by the formula quoted in Chapter 4 (p. 105), or may easily be calculated in steps as in the following example:

EXAMPLE
Calculate the amount of compressive reinforcement required for the ultimate limit state if the imposed load on the rectangular R.C. section in the previous example is to be increased to 20 kN/m (1370 lbf/ft) without increasing the dimensions of the section.

Total design load required at ultimate limit state

$$w_{ud \text{ req.}} = 1.4(17 + 2) + 1.6 \times 20 = 58.6 \text{ kN/m (4010 lbf/ft)},$$

$$M_{ud \text{ req.}} = 58.6 \times 25^2 \times 0.125 = 4570 \text{ kN m (40 400 000 lbf in)}.$$

Maximum value of M_{ud} without compressive reinforcement

$0.15 f_{cu} bd^2 = 0.15 \times 45 \times 600 \times 1000^2 = 4050$ kN m (35 800 000 lbf in).

Moment to be provided by compressive reinforcement

$\Delta M_{ud\ req.} = 4570 - 4050 = 520$ kN m (4 600 000 lbf in).

Stress in compressive reinforcement

$$= \frac{2000}{2300 + 420} \times 420 = 309 \text{ N/mm}^2 \text{ (44 900 lbf/in}^2\text{)}.$$

Assume distance of compressive reinforcement from top,

$d' = 50$ mm (2 in), and therefore $d - d' = 1000 - 50$
$$= 950 \text{ mm (37.4 in)}.$$

Area of compressive reinforcement,

$$A_s' = \frac{520 \times 10^6}{309 \times 950} = 1770 \text{ mm}^2 \text{ (2.74 in}^2\text{)}.$$

3–30 mm bars give 2170 mm^2 (3–1¼ in bars give 3.7 in^2).

Area of tensile reinforcement,

$$A_s = \frac{0.15 f_{cu} bd^2}{0.87 f_y \times 0.75d} + \frac{\Delta M_{ud\ req.}}{0.87 f_y (d - d')}$$

$$= \frac{4050 \times 10^6}{0.87 \times 420 \times 0.75 \times 1000}$$

$$+ \frac{520 \times 10^6}{0.87 \times 420 \times 950} = 16\ 300 \text{ mm}^2 \text{ (25.2 in}^2\text{)}.$$

23-30 mm bars give 16 300 mm^2 (21-1¼ in bars give 25.8 in^2), i.e. 2 additional bars are required.

Chart for tensile and compressive reinforcement

Figure 7.4 is a graphical presentation of the British Code formulae for the moment of resistance of reinforced concrete sections with and without compressive reinforcement.

The curves have been related to the overall dimensions, b and h, and drawn for various values of d/h; this indicates the influence of different amounts of cover and is consistent with charts for compression and bending in which a similar arrangement is generally adopted.

EXAMPLE

In the previous example $d/h = 1000/1200 = 0.83$
and

$$\frac{M_{ud\ req}}{f_{cu}\ bh^2} = \frac{4570 \times 10^6}{45 \times 600 \times 1200^2} \simeq 0.12.$$

The reinforcement ratios are found from Fig. 7.3 to be

$$\frac{f_y A_s}{f_{cu}\ bh} \simeq 0.21 \quad \text{and} \quad \frac{f_y A_s'}{f_{cu}\ bh} = 0.025.$$

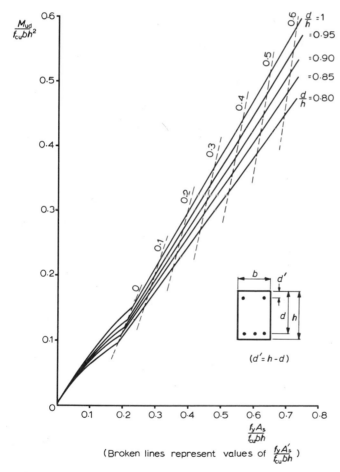

Figure 7.4   Design of chart for rectangular R.C. sections

The areas of reinforcement are therefore

$$A_s = \frac{0.21 \times 45 \times 600 \times 1200}{420} = 16\ 200\ \text{mm}^2\ (25.0\ \text{in}^2),$$

$$A_s' = \frac{0.025 \times 45 \times 600 \times 1200}{420} = 1\ 900\ \text{mm}^2\ (2.94\ \text{in}^2).$$

These results correspond fairly closely to the more accurate ones obtained by the more lengthy calculations.

The chart gives a rapid indication of the load capacity of a section with compressive reinforcement and is therefore useful for preliminary dimensioning, as in the following example.

EXAMPLE

Investigate in the previous example the minimum size of rectangular R.C. section that can be designed, using the maximum compressive reinforcement.

The maximum area of compressive reinforcement permissible under the British Code is 0.04 bd. Therefore allowing for a ratio $d/h = 0.80$,

$$\frac{f_y A_s'}{f_{cu} bh} = \frac{420}{45} \times 0.04 \times 0.80 = 0.30.$$

From Fig. 7.4 the corresponding value of $M_{ud}/f_{cu} bh^2$ is 0.27. Since $M_{ud} = 4000$ kN m (36 000 000 lbf in) the value of h corresponding to $b = 600$ m (23.6 in) is found to be 740 mm (29.2 in). The required areas of reinforcement are

$$A_s = 20\ 900\ \text{mm}^2\ (32.4\ \text{in}^2): 30\text{--}30\ \text{mm bars},$$
$$A_s' = 14\ 300\ \text{mm}^2\ (22.1\ \text{in}^2): 21\text{--}30\ \text{mm bars}.$$

Alternatively the overall depth of 1200 mm (47.2 in) required for a beam without compressive reinforcement, may be adopted and the breadth reduced. b is then found to be 229, say 230 mm (9 in), and the reinforcement is

$$A_s = 13\ 000\ \text{mm}^2\ (20.2\ \text{in}^2): 19\text{-}30\ \text{mm bars},$$
$$A_s' = 11\ 100\ \text{mm}^2\ (17.2\ \text{in}^2): 16\text{-}30\ \text{mm bars}.$$

The two solutions are shown in Fig. 7.5. The first involves a very large span/depth ratio (25/0.740 = 33.8) and the limit state of deflexion must be checked. Using the formula in Table 2.2, the neutral axis depth ratio, ξ, is found to be 0.487, and $I_c = 37\ 400 \times 10^6\ \text{mm}^4$ (899 000 in^4). The instantaneous deflexion is 125 mm (4.9 in) and the long term deflexion 188 mm (7.4 in); thus

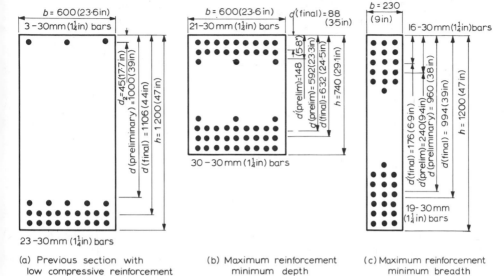

(a) Previous section with
low compressive reinforcement

(b) Maximum reinforcement
minimum depth

(c) Maximum reinforcement
minimum breadth

Figure 7.5   Cross sections of R.C. beams with compressive reinforcement

although the former is greater than that of the beam without compressive reinforcement, the reduced factor for the time effect results in a lower long-term deflexion.

The second alternative is much more economical in reinforcement but is a very slender beam and the ratios

$$\frac{l}{b} = \frac{25}{0.23} = 109 \quad \text{and} \quad \frac{ld}{b^2} = \frac{25 \times 0.8 \times 1.2}{(0.23)^2} = 454$$

are well outside the limits normally specified.

It could therefore only be used with lateral support to ensure stability.

Flanged prestressed sections — serviceability limit state

It has already been explained how the dimensions of the compression flange (bh_f) and the web or rib (b_w) may be selected to satisfy the ultimate limit state. The dimensions of the bottom flange (b_{inf}, h_{inf}) must be such that the section modulus is sufficient, this being determined by the serviceability limit state.

For prestressed concrete the serviceability limit state is defined in the British Code by allowable stresses in the concrete. Since these are largely the same as those in the earlier Code CP 115 : 1959 and, as will be recalled, the partial load factor (γ_f) for dead and imposed loads is unity, the recommendations really amount to the retention of the permissible stress method.

The allowable compressive stresses are given in Table 7.1. They are specified to safeguard against spalling of the concrete, although it could be argued that this is not likely to take place much before general compressive failure, so that if the compressive zone of a member is adequate for the ultimate limit state, these allowable stresses are redundant.

The allowable tensile stress has an important influence on the size of the bottom flange. It is chosen with regard to the extent to which cracking is to be permitted, according to the three classes of prestressed member defined in Chapter 6.

In Class 1 no cracking is permitted. This class is specified for members exposed to a corrosive environment, for liquid-retaining structures and for members subject to dynamic loading where the fatigue of steel in a cracked member would constitute a danger. The limit state is defined by the decompression load, at which the prestress is exactly cancelled by the bending stress at the tensile face, i.e. the permissible tensile stress is zero. Under these conditions, even if cracks were to be caused by accident, they would remain closed for values of the design load up to the serviceability limit state. At transfer a small tensile stress may be allowed where it is due solely to the prestressing force and a value of 1 N/mm^2 (145 lbf/in^2) is recommended.

In Class 2 the design load for the serviceability limit state corresponds to an allowable tensile stress less than the tensile stress at which cracking occurs. In the British Code the allowable flexural tensile stress $f_{t \text{ adm}}$ at the design load is given by the following formulae:

Members with pre-tensioned steel ($f_{cu} \geqslant 40 \text{N/mm}^2$ (5800 lbf/in^2))

$$f_{t \text{ adm}} = 1.9 + 0.027 f_{cu} \text{N/mm}^2 \ (275 + 0.027 f_{cu} \text{ lbf/in}^2)$$

Members with post-tensioned steel ($f_{cu} \geqslant 30 \text{ N/mm}^2$ (4350 lbf/in^2))

$$f_{t \text{ adm}} = 1.2 + 0.027 f_{cu} \text{ N/mm}^2 \ (175 + 0.027 f_{cu} \text{ lbf/in}^2)$$

Table 7.1   *Allowable compressive stresses in concrete at serviceability limit state (British Code)*

		Bending (or approximately triangular distribution of stress)	Axial load (or approximately uniform distribution of stress)
At initial transfer of prestress	$f_{c \text{ p adm}}$	$0.5 f_{ci}$	$0.4 f_{ci}$
At design load	$f_{c \text{ adm}}$	$0.33 f_{cu}$	$0.25 f_{cu}$

f_{ci} = characteristic compressive strength of concrete at initial transfer of prestress.

f_{cu} = characteristic compressive strength of concrete at 28 days or at time of first applying design load, whichever is less.

The increase of 0.7 n/mm^2 (100 lbf/in^2) in the tensile stress allowed with pre-tensioned steel takes advantage of the improved bond and crack control characteristics of pre-tensioned wires over post-tensioned cables, provided that they are well-distributed in the section.

The values given by the above formulae may be increased by up to 1.7 N/mm^2 (250 lbf/in^2) provided that the allowable stress does not exceed three-quarters of the experimentally established stress at which cracking appears. Good crack control must be ensured by distribution of pre-tensioned wires or by well distributed supplementary reinforcement. It is also necessary to ensure that any cracks which may occur under an accidental high load should close again on reverting to normal conditions; it is therefore specified that when the allowable tensile stress is increased in this way the prestress in the concrete (presumably at the tensile face) must be at least 10 N/mm^2 (1450 lbf/in^2). A still further increase of tensile stress is permitted for temporary loads which are high in relation to the normal load, provided that the stress is compressive under the latter conditions to ensure closure of cracks; it is in fact desirable that the concrete stress under dead load should be compressive in the majority of Class 2 members.

The allowable flexural tensile stress at transfer $f_{t\,p\,adm}$ is given by the above formulae, using the cube strength at transfer, in place of the characteristic cube strength at 28 days, or at the age at which the design load is first applied.

Class 3 prestressed members are designed to permit permanent flexural cracks of restricted width. Their behaviour is therefore essentially similar to that of conventional reinforced concrete (they are sometimes known as prestressed reinforced concrete) and their advantage would appear to lie in economy of material rather than in improved structural properties. The crack width may be checked by a formula based on the stress in the reinforcement, as for reinforced concrete, or may be related to the stress which would occur at the tensile face of the concrete in the absence of cracking. The writer believes the former method to be in principle the more accurate of the two, but the latter, which is at present adopted in the British Code, is somewhat simpler to apply as it does not require calculation of the stress in the reinforcement, and the method of calculation is basically the same as for Classes 1 and 2. The allowable 'hypothetical' flexural tensile stresses have already been given in Table 2.4 (p. 66).

Flanged prestressed sections — stress conditions and minimum dimensions

A prestressed member must be designed so that the stresses in the concrete will be less than the specified allowable values at all points in the section, at any possible load up to the design load for the serviceability limit state and with any possible value of the prestress between the initial maximum value and the value after all losses have occurred.

The stress considered here is the resultant of the prestress and the stress due to the load, either of which may individually be greater than the allowable value.

The problem can be defined by four stress relationships, established for the two extreme faces of the section at two critical combinations of prestress and load. These critical combinations are

(I) The maximum prestressing force P, at transfer, in combination with the minimum moment M_{min} which is sustained throughout the life of the member.

(II) The minimum prestressing force ηP, after all losses, in combination with the total design moment M_d for the serviceability limit state.

The four relationships are shown on the stress diagrams in Fig. 7.6. At the bottom of a beam subject to positive bending (bending stress tensile at bottom) the maximum compressive stress will occur under condition (I) and the maximum tensile stress under condition (II). The stress conditions are therefore

$$f_{inf} - \frac{M_{min}}{Z_{inf}} < f_{c\,p\,adm} \qquad\qquad 7.1$$

$$-\eta f_{inf} + \frac{M_d}{Z_{inf}} < f_{t\,adm} \qquad\qquad 7.2$$

f_{inf} is the compressive prestress in the concrete at the bottom at transfer and the permissible stresses $f_{c\,p\,adm}$ and $f_{t\,adm}$ are positive in sign.

At the top the maximum tensile stress will occur under condition (I) and the maximum compressive stress will occur under condition (II), when there is a tensile prestress at the top, as in the majority of instances. Hence

$$f_{sup} - \frac{M_{min}}{Z_{inf}} < f_{c\,p\,adm}, \qquad\qquad 7.3$$

$$-\eta f_{sup} + \frac{M_d}{Z_{sup}} \leqslant f_{c\,adm}. \qquad\qquad 7.4$$

f_{sup} is the tensile prestress in the concrete at the top at transfer. If there is a compressive prestress the stress condition 7.4 is slightly modified, as the resultant compressive stress will be a maximum if the total moment occurs before there have been any losses of prestress.

$$-f_{sup} + \frac{M_d}{Z_{sup}} \leqslant f_{c\,adm}. \qquad\qquad 7.5$$

Since f_{sup} is defined as a tensile stress the sign of the numerical value of $-f_{sup}$ will be positive, as will be clear from the diagram in Fig. 7.6.

A design formula for the required minimum section modulus for the bottom of the section is derived by eliminating f_{inf} from conditions 7.1 and 7.2.

$$Z_{inf} \leqslant \frac{M_d - \eta M_{min}}{\eta f_{c\,p\,adm} + f_{t\,adm}}.$$

The allowable stresses $f_{c\,p\,adm}$ and $f_{t\,adm}$ are specified. The ratio of loss of prestress, η, usually lies in the range of 0.75 to 0.80 for members with pre-tensioned tendons and 0.80 to 0.85 for post-tensioning. After completion of the design the ratio may, if desired be more accurately assessed by the methods described in Chapter 3.

The total moment M_d is the value calculated suing the factors γ_f given in the previous chapter for the serviceability limit state, and it should be noted that M_{min} is included in this total moment. The minimum moment is frequently the moment due to the self weight of the beam, but a lower value may occur during construction, as for example in the lifting of a precast beam at points other than the supports; however, some concessions may be allowed in the permissible stresses for such loadings of short duration. Dead load applied after prestressing cannot be included unless a second stage of prestressing is subsequently carried out, a technique occasionally used for beams supporting very large dead loads. It

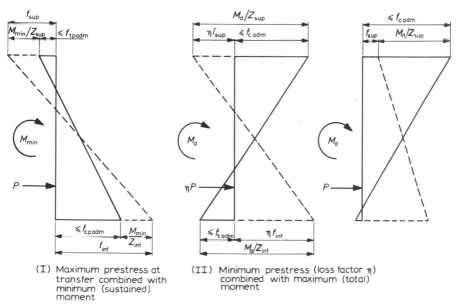

(I) Maximum prestress at transfer combined with minimum (sustained) moment

(II) Minimum prestress (loss factor η) combined with maximum (total) moment

Figure 7.6   Stress conditions in a prestressed beam

will be seen from the formula that because of the loss of prestress the sustained moment has some influence on the required size of section, and it is incorrect to state that the latter is influenced only by the live load or that the self weight is entirely counteracted by the prestress.

A corresponding although less important formula may be derived for the section modulus for the top of the section from the stress conditions 7.3 and 7.4 or 7.3 and 7.5

$$Z_{sup} \geqslant \frac{M_d - \eta M_{min}}{f_{c\ adm} + \eta f_{t\ p\ adm}}.$$

The next step in dimensioning a flanged prestressed concrete section is to determine the breadth b_{inf} and depth h_{inf} of the bottom flange so that with the values of b, d, h_f and b_w already selected, Z_{inf} is equal to or slightly greater than the minimum given by the above formula. When this is done the value of Z_{sup} will usually be found to be adequate. This may be done by trial and error, but the work entailed in the calculation of several sets of dimensional properties is quite considerable, even if the section is approximated by three rectangles as in Fig. 7.7. The operation can therefore be shortened appreciably by the use of the tables given as Appendix B in which the ratios of the section moduli to those of the circumscribing rectangle, $Z_{inf}/(bh^2/6)$ and $Z_{sup}/(bh^2/6)$ are given for a range of values of the ratios of the each flange dimensions to the overall depth or breadth. This is illustrated in the continuation of the example already commenced.

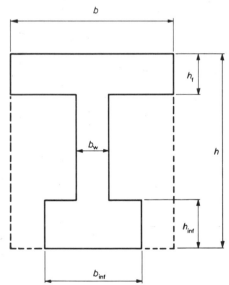

Figure 7.7   Simplified flanged section for approximate calculation of dimensional properties

EXAMPLE

Suitable allowable stresses would be

$f_{a\ adm}$ $= 0.33 \times 45 = 15.0 \text{ N/mm}^2$ (2170 lbf/in^2).

$f_{c\ p\ adm} = 0.5 \times 37.5 = 18.75 \text{ N/mm}^2$ (2700 lbf/in^2).

$f_{t\ adm}$ $= 1.2 \times 0.027 \times 45 = 2.4 \text{ N/mm}^2$ (350 lbf/in^2)(for post-tensioning).

$f_{t\ p\ adm} = 1.2 \times 0.027 \times 37.5 = 2.2 \text{ N/mm}^2$ (320 lbf/in^2)(for post-tensioning).

Since there is a 15% loss of prestress after transfer, the loss ratio $\eta = 1 - 0.15 = 0.85$.

The sustained load is in this instance the self-weight of the beam which has already been estimated (p. 177) and the total load for the serviceability limit state is

$$w_d = 9.1 \times 1.0 + 2 \times 1.0 + 15 \times 1.0 = 26.1 \text{ kN/m (1780 lbf/ft)}.$$

Hence

$$M_{min} = 9.1 \times 25^2 \times 0.125 = 710 \text{ kN/m (6 220 000 lbf in)}.$$

$$M_d = 26.1 \times 25^2 \times 0.125 = 2040 \text{ kN/m (18 050 000 lbf in)}.$$

The minimum section moduli required are

$$Z_{inf} = \frac{2040 - 0.85 \times 710}{0.85 \times 18.75 \times 2.4} \times 10^6 = 78.5 \times 10^6 \text{ mm}^3 \text{ (4780 in}^3\text{)},$$

$$Z_{sup} = \frac{2040 - 0.85 \times 710}{15 + 0.85 \times 2.2} \times 10^6 = 85.0 \times 10^6 \text{ mm}^3 \text{ (5170 in}^3\text{)}.$$

The section modulus ratios required are

$$\frac{Z_{inf}}{bh^2/6} = \frac{78.5 \times 10^6}{600 \times 1300^2/6} = 0.465,$$

$$\frac{Z_{sup}}{bh^2/6} = \frac{85.0 \times 10^6}{600 \times 1300^2/6} = 0.505.$$

Two of the dimensional ratios have already been fixed, namely

$$\frac{b_w}{b} = \frac{150}{600} = 0.25 \quad \text{and} \quad \frac{h_f}{h} = \frac{230}{1300} = 0.18.$$

The table of section modulus ratios (Appendix B) enables alternative solutions to be examined. Here for example it is seen that a symmetrical I section $((b_{inf})/(b) = 1, (h_{inf})/(h_{inf})/(h) \approx 0.15)$ has section modulus ratios

$$\frac{Z_{inf}}{bh^2/6} = \frac{Z_{sup}}{bh^2/6} = 0.777,$$

and is much stiffer than necessary; the beam may therefore be made lighter by the adoption of a narrow bottom flange. It is not, however, necessarily most economical to design a section which only just meets the minimum requirements, as a larger section will often require less prestress. The ratio $Z_{sup}(bh^2/6)$ will almost always be greater than the minimum indicated by the formula, as it is largely controlled by the dimensions of the top flange, which have been determined for the ultimate limit state; moreover it is often necessary for it to be greater than the minimum for a further reason which will become evident when calculating the position of the tendons.

In this instance a bottom flange of breadth 300 mm (12 in) and average depth 200 mm (8 in) will be adopted (Fig. 7.8). From the table,

putting $(b_{inf})/(b) = (300)/(600) = 0.50$, $(h_{inf})/(h) = (200)/(1300) \simeq 0.15$ and interpolating for $h_f = 0.18$ the section modulus ratios are found to be 0.521 and 0.712 and the approximate values of the section moduli are

$$Z_{inf} = 88 \times 10^6 \text{ mm}^3 \text{ (5350 in}^3\text{)}$$
$$Z_{sup} = 121 \times 10^6 \text{ mm}^3 \text{ (7370 in}^3\text{)}.$$

The cross sectional area can be rapidly calculated and is found to be 328 500 mm² (510 in²). The weight of the section is thus about 7.0 kN/m (480 lbf/ft) which is less than the value of 9.1 kN/m (625 lbf/ft) originally allowed.

For subsequent calculations the height of the centroid (y_{inf}) and the radius of gyration (i) of the section are required.

$$y_{inf} = \frac{Z_{sup}h}{Z_{inf} + Z_{sup}} = \frac{0.712 \times 1300}{0.521 + 0.712} = 755 \text{ mm (30 in).}$$

$$I = Z_{inf}y_{inf} = 88 \times 755 = 66\,500 \times 10^6 \text{ mm}^4 \text{ (160 000 in}^4\text{)}.$$
$$i^2 = 66\,500 \times 10^6/328\,500 = 20\,200 \text{ mm}^2 \text{ (313 in}^2\text{)}.$$

Prestressing tendons

In order to calculate the total prestressing force required and its position, one must first find the necessary prestress in the concrete. If the minimum size of section were to be used, so that the section moduli Z_{inf} and Z_{sup} were exactly equal to the minimum values indicated by the design formulae given above, the prestress would be uniquely determined both for the bottom and the top of the section. Usually, however, the section is somewhat greater than the minimum, with the result that the prestress can lie between an upper and a lower limit, these limits being defined at the bottom of the section by the above stress conditions 7.1 and 7.2 and at the top of the section by the conditions 7.3 and 7.4, using the values of Z_{inf} and Z_{sup} for the section as actually designed.

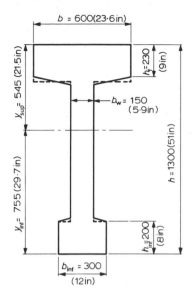

Figure 7.8　Dimensioning of cross section of prestressed beams for ultimate and serviceability limit states

Compressive pre-stress at bottom

$$f_{\text{inf}} \leqslant \frac{M_{\text{min}}}{Z_{\text{inf}}} + f_{\text{c p adm}} , \qquad 7.1'$$

$$f_{\text{inf}} \geqslant \frac{1}{\eta} \left(\frac{M_{\text{d}}}{Z_{\text{inf}}} - f_{\text{t adm}} \right) \qquad 7.2'$$

Tensile pre-stress at top

$$f_{\text{sup}} \leqslant \frac{M_{\text{min}}}{Z_{\text{sup}}} + f_{\text{t p adm}} , \qquad 7.3'$$

$$f_{\text{sup}} \geqslant \frac{1}{\eta} \left(\frac{M_{\text{d}}}{Z_{\text{sup}}} - f_{\text{c adm}} \right) . \qquad 7.4'$$

Any value of the prestress within these limits may be used, but it should be noted that the lowest prestressing force will be obtained by selecting the minimum compressive prestress, indicated by 7.2', at the bottom, and the maximum tensile prestress, indicated by 7.3', at the top.

Having decided on the prestress in the concrete, the position and magnitude of the corresponding prestressing force must be found. From the formulae given in Chapter 1 (p. 14) for the prestress at the bottom and top of the section the distance e of the prestressing force below the centroidal axis is

$$e = \frac{i^2 \left(f_{\text{inf}} + f_{\text{sup}} \right)}{h \left\{ f_{\text{inf}} - \dfrac{y_{\text{inf}}}{h} \left(f_{\text{inf}} + f_{\text{sup}} \right) \right\}}$$

Since f_{inf} is a compressive stress and f_{sup} a tensile stress, the term $(f_{inf} + f_{sup})$ is the algebraic difference of the prestress at the bottom and top of the section.

Sometimes, however, the position of the prestressing force will be controlled by practical limitations. This will occur when the minimum moment (M_{min}) is relatively large and the theoretical value of e indicates a position which is either below the soffit of the section or too low for tendons to be placed with adequate cover. Then e must be adjusted to bring the prestressing force to a suitable position within the section, e.g. near the centre of the bottom flange.

The magnitude of the prestressing force (P) is next calculated from the selected value of e so that the prestress at the bottom has the required value, f_{inf}.

$$P = 1 + \frac{e\, y_{inf}}{i^2}.$$

The effect of adjusting the eccentricity of the force to a lower value will be to decrease the tensile prestress at the top, as shown in Fig. 7.9. The resultant compressive stress under the total moment (M_d) will therefore be greater, and it is therefore necessary to check that the allowable stress is not exceeded, using the relation

$$\frac{M_d}{Z_{sup}} - \frac{\eta P}{A}\left(\frac{e\, y_{sup}}{i^2} - 1\right) \leqslant f_{c\,adm}.$$

If the sign of the bracketed expression is negative, the prestress at the top is compressive and η should be put equal to unity.

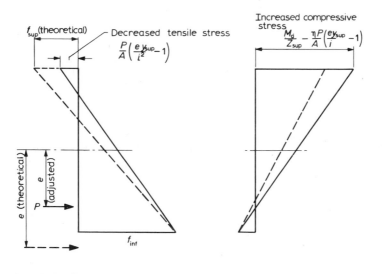

(a) Maximum prestress only   (b) Minimum prestress
                                            + total moment

Figure 7.9   Adjustment of position of prestressing force

For the above condition to be fulfilled after decreasing the eccentricity of the prestressing force, the value of Z_{sup} must be greater than the minimum indicated by the design formula. It has been seen that this is usually so when the top flange has been dimensioned for the ultimate limit state. Two alternative graphical methods which may be used to determine suitable values of P and e are given in Appendix C. The first of these was developed by Magnel [7.1], the second by the author [7.2].

The stress conditions may be satisfied by any system of tendons in which the required total force P acts at the required eccentricity e, but the actual arrangement must take account of other considerations. Where pre-tensioned tendons are used, one row should be placed as near as possible to the bottom face and another close to the top face for crack control. The local stresses at the ends of the member will be less severe if the remaining tendons are sufficiently distributed to avoid large concentrated forces. Spacing is governed by the maximum size of aggregate and the need for the concrete to flow easily to all parts of the section.

Post-tensioned tendons cannot be so easily distributed and crack control is achieved by the use of supplementary reinforcement. The tendons may have to be located in such a way that they can be curved upwards into the web to improve the stress conditions and resistance to shear near the ends. Cover and spacing must take into account both the placing of the concrete and the resistance to bursting when the ducts are pressure grouted.

EXAMPLE
The limiting values of the prestress required in the section already designed, are first obtained.

$$\frac{1}{0.85}\left(\frac{2040}{88} - 2.4\right) \leqslant f_{inf} \leqslant \frac{710}{88} + 18.75,$$

i.e.

$$25.70 \leqslant f_{inf} \leqslant 26.80 \text{ N/mm}^2.$$

From 7.3' and 7.4',

$$\frac{1}{0.85}\left(\frac{2040}{121} - 15.0\right) \leqslant f_{sup} \leqslant \frac{710}{121} + 2.2,$$

i.e.

$$2.18 \leqslant f_{sup} \leqslant 8.07 \text{ N/mm}^2.$$

Selecting $f_{inf} = 26$ N/mm² (3770 lbf/in²) and $f_{sup} = 8$ N/mm² (1160 lbf/in²) (tensile), the theoretical eccentricity is found to be

$$e = \frac{202\ 000\ (26 + 8)}{1300\ [26 - (26 + 8) \times 0.580]} = 837 \text{ mm (32.8 in).}$$

This is greater than y_{inf} and is therefore impracticable; the value of e is therefore adjusted to 650 mm (25.6 in) assuming that the average position of the tendons will be at the mid depth of the bottom flange.

Then

$$P = \frac{26 \times 328\,500}{1 + \dfrac{650 \times 755}{202\,000}} = 2500 \text{ kN (563 000 lbf).}$$

The compressive stress at the top of the section under the total moment is

$$\frac{2040}{121} - \frac{0.85 \times 2500 \times 10^3}{328\,500} \left(\frac{650 \times 545}{202\,000} - 1 \right) = 12.01 \text{ N/mm}^2 \text{ (1745 lbf/in}^2\text{).}$$

This is less than the permissible stress of 15 N/mm² (2170 lbf/in²).

A possible prestressing arrangement using cables of 7 mm (0.276 in) wires is shown in Fig. 7.10. There are 5 cables of 12 wires in 50 mm (2 in) ducts. If the prestress in the steel at transfer is 1050 N/mm² (152 000 lbf/in²), i.e. 70% of the tensile strength, the total prestressing force will be 2430 kN (547 000 lbf) and the resultant eccentricity for the given cable positions will be 655 mm

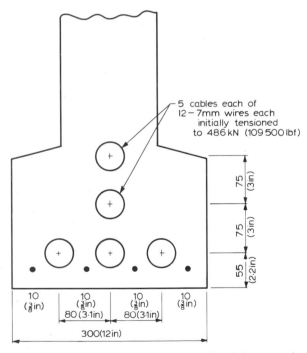

Figure 7.10   Detail of tendons and tensile reinforcement in cross section of prestressed beam

(25.8 in), giving a prestress of 25.5 N/mm² (3740 lbf/in²) close to the lower limit calculated above.

The area of prestressed steel is 2310 mm² (3.58 in²) at an effective depth of 1200 mm (47.3 in). The total area of steel calculated for the ultimate limit state was 2350 mm² at an estimated effective depth of only 1130 mm (44.5 in), so that no supplementary reinforcement is required in this example. It is, however, advisable to place a small amount of reinforcement close to the bottom and top of the section to control possible cracking, especially shrinkage cracking before the beam is prestressed, and this has been shown in Fig. 7.10.

Dimensioning of remainder of member

The calculations have so far concerned one section of the member, and although much of the work of dimensioning is concerned with a critical section, it is necessary throughout to have the remainder of the member in mind. This may differ from the critical section in three principal ways:

(a) Differing cross section.
(b) Variation of the amount of reinforcement.
(c) Variation in the position, and occasionally in the magnitude of the prestressing force.

Suitable leading dimensions may be established for any section by the method described. The geometry of members of varying cross section is to some extent predetermined, usually by the type of profile required, so that when the dimensions of the critical section are selected, those of the remaining sections are also fixed. In a member of varying cross section, however, the critical section does not always occur at the point of maximum moment as in a prismatic member; for example in a simply supported beam of 'hog-backed' or 'fish-bellied' profile it is often nearer to the 3/8 span points. The amount of flexural reinforcement required at the critical section is often more than is necessary in other parts of the member, and the 'cut-off' points of the reinforcing bars may easily be found. For a prismatic member the contribution of the tensile bars is marked on the bending moment diagram by a series of horizontal lines which intersect the bending moment curve at the point at which each bar becomes unnecessary. The bar must, however, be continued for a further anchorage length, sufficient to develop the required tensile stress at this point.

EXAMPLE

The bending moment diagram (ultimate limit state) for the reinforced concrete beam in the former example is shown in Fig. 7.11. At the critical section at

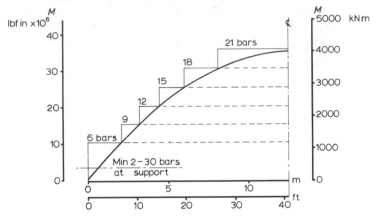

Figure 7.11   Cut-off diagram for tensile reinforcing bars

midspan the moment is 4 000 kN m (35 400 000 lbf in) and the total moment of resistance provided by the 21-30 mm (1¼ in) bars is about 4080 kN m (36 000 000 lbf in), which is divided on the diagram into increments of 580 kN m (5 120 000 lbf in) to show the cut-off points for groups of 3 bars.

The anchorage length of each bar beyond the cut-off point is based on an average bond stress as described in Chapter 2 (p. 54ff). According to the British Code the anchorage bond stress for bars of high bond strength in tension, in concrete of cube strength 45 N/mm² (495 lbf/in²). The anchorage length is therefore

$$\frac{0.87 \times 420}{4 \times 2.5} = 36 \text{ diameters} = 1080 \text{ mm (42 in)}$$

which can be reduced by the appropriate allowance if a hook is used.

In view of the anchorage length required it is necessary to continue the last two groups of 3 bars to the end of the beam. Actually the minimum reinforcement required at the support, where the bending moment is zero, is

$$0.15\% \, bd = \frac{0.15}{100} \times 600 \times 1112 = 1000 \text{ mm}^2 \text{ (1.56 in}^2\text{)}.$$

This is provided by two 30 mm bars.

The shear resistance has to be considered in conjunction with the termination of tensile reinforcement, as it is affected both by the amount of tensile reinforcement and by the use of bent-up bars as shear reinforcement. This aspect will be discussed in Chapter 9.

In a prestressed member, at a section subject to a large minimum moment, the eccentricity of the prestressing force has to be large to counteract the bending stress due to the minimum moment. The effect of reducing the minimum moment without altering the prestress would be to overstress the concrete both in compression and tension. This can be avoided by moving the prestressing force towards the centroid of the section so that the extreme compressive and tensile values of the prestress at the outer limits of the section are reduced. This practice is exemplified by the use of 'draped' post-tensioned tendons, which in a simply supported beam are located in the lower part of the section at midspan and curve upwards towards each end of the beam. Pre-tensioned tendons may be 'harped', i.e. deflected into a varying profile consisting of a number of straight lengths by the use of deviating devices in the mould before casting.

If the magnitude and position of the prestressing force is determined for the critical section by the method described it is possible to establish the upper and lower limits of this force at all other points along the member. These limits are defined by four equations formed by combining the stress conditions 7.1 to 7.4 on page 192 with the prestress equations for the bottom and top of the section, namely

$$f_{\text{inf}} = \frac{P}{A}\left(1 + \frac{Ae}{Z_{\text{inf}}}\right)$$

$$f_{\text{sup}} = \frac{P}{A}\left(\frac{Ae}{Z_{\text{sup}}} - 1\right)$$

This gives

$$e \leqslant \frac{Z_{\text{inf}}\, f_{\text{c p adm}}}{P} - \frac{Z_{\text{inf}}}{A} + \frac{M_{\text{min}}}{P}, \tag{7.1(a)}$$

$$e \geqslant \frac{Z_{\text{inf}}\, f_{\text{t adm}}}{\eta P} - \frac{Z_{\text{inf}}}{A} + \frac{M_{\text{d}}}{\eta P}, \tag{7.2(a)}$$

$$e \leqslant \frac{Z_{\text{sup}}\, f_{\text{t adm}}}{P} + \frac{Z_{\text{sup}}}{A} + \frac{M_{\text{min}}}{P}, \tag{7.3(a)}$$

$$e \geqslant \frac{-Z_{\text{sup}}\, f_{\text{t p adm}}}{\eta P} + \frac{Z_{\text{sup}}}{A} + \frac{M_{\text{d}}}{\eta P}. \tag{7.4(a)}$$

The sign convention for e is positive when the prestressing force lies below the centroid of the section; sagging moments are positive as also are all permissible stresses (f_{adm}) whether tensile or compressive. Clearly only two of the above four equations will define the limits of e at any given section.

EXAMPLE

The equations 7.1(a)–7.4(a) are used to plot the limits within which the prestressing force, designed for the section of the beam at midspan, must lie.

$$e \leqslant \frac{88 \times 10^6 \times 18.75}{2430 \times 10^3} - \frac{88 \times 10^6}{328\ 500} + \frac{M_{min}}{2430 \times 10^3}$$

$$\leqslant 412 + \frac{M_{min}}{2430 \times 10^3}\ \text{mm.} \qquad\qquad 7.1\text{(a)}$$

$$e \geqslant -\frac{88 \times 10^6 \times 2.4}{0.85 \times 2430 \times 10^3} - \frac{88 \times 10^6}{328\ 500} + \frac{M_d}{0.85 \times 2430 \times 10^3}$$

$$\geqslant -370 + \frac{M_d}{2060 \times 10^3}\ \text{mm.} \qquad\qquad 7.2\text{(a)}$$

$$e \leqslant \frac{121 \times 10^6 \times 2.2}{2430 \times 10^3} + \frac{121 \times 10^6}{328\ 500} + \frac{M_{min}}{2430 \times 10^3}$$

$$\leqslant 479 + \frac{M_{min}}{2430 \times 10^3}\ \text{mm.} \qquad\qquad 7.3\text{(a)}$$

$$e \geqslant -\frac{121 \times 10^6 \times 15.0}{0.85 \times 2430 \times 10^3} + \frac{121 \times 10^6}{328\ 500} + \frac{M_d}{0.85 \times 2430 \times 10^3}$$

$$\geqslant -509 + \frac{M_d}{2060 \times 10^3}\ \text{mm.} \qquad\qquad 7.4\text{(a)}$$

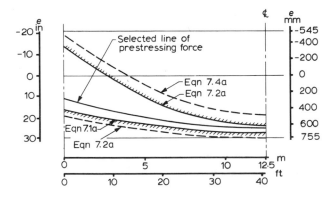

Figure 7.12   Tendon profile diagram

The curves represented by the four equations are shown in Fig. 7.12.

In all prismatic members with a constant prestressing force the curves of the equations 7.1(a) and 7.3(a) are parallel as also are the curves of 7.2(a) and 7.4(a). This reduces the amount of work, as it may quickly be seen which two curves define the inner limits; in the present example these are the curves of 7.1(a) and 7.2(a). If, however, there is a member of varying cross section, or if the prestressing force is varied by terminating some tendons before the end of the member, the limits may be defined by different equations at different points, i.e. the upper or lower limits curves may cross.

The profiles of the tendons are largely controlled by the need for suitable spacing at the end anchorages in order to distribute the large concentrated forces involved, which will be discussed further in Chapter 9. If the anchorages are arranged as in Fig. 9.14 and each tendon is draped in a parabolic curve, the line of the resultant prestressing force is as shown in Fig. 7.12, lying within the necessary limits.

Water-retaining structures in reinforced concrete

The design conditions for water retaining reinforced concrete structures differ from those of other types, in that cracking has to be avoided in order to ensure impermeability. The serviceability limit state is therefore the point at which cracking first occurs and this part of the calculation is made for the uncracked section.

The leading dimensions of a flexural member, such as the vertical cantilever wall of a tank, designed for these conditions, are jointly controlled by the ultimate and serviceability limit state which have to be considered together. A maximum tensile stress in the concrete is specified for the latter limit state. If M_{ud} is the moment at the limit state of collapse

$$M_{ud} = (0.87 f_y) A_s z,$$

$$\frac{M_{ud}}{bh^2} = (0.87 f_y) \frac{A_s}{bh} \cdot \frac{z}{d},$$

where

f_y = characteristic yield stress of reinforcement,

A_s = area of tensile reinforcement,

b = breadth of section,

h = overall depth of section,

z = lever arm.

Assume that the centroidal axis is at the mid-depth of the uncracked section, and the tensile and compressive reinforcement are equidistant from the axis. Then the second moment of area is given by

$$\frac{bh^3}{12} + (A_s + A_s')(\alpha - 1)\left(d - \frac{h}{2}\right)^2,$$

where d = effective depth of tensile reinforcement,

A_s' = area of compressive reinforcement,

α = modular ratio.

If M_d is the moment and $f_{t\,adm}$ the flexural tensile stress in the concrete at the serviceability limit state,

$$f_{t\,adm} = \frac{M_d}{I} \cdot \frac{h}{2},$$

$$\frac{M_d}{bh^2} = \frac{f_{t\,adm}}{6}\left\{1 + 3\,(\alpha - 1)\left(\frac{A_s + A_s'}{bh}\right)\left(2\frac{d}{h} - 1\right)^2\right\}.$$

Substituting for A_s/bh from the equation for M_{ud}

$$\frac{M}{bh^2} = \frac{f_{t\,adm}}{6 - \dfrac{3(\alpha - 1)}{z/h}\left(\dfrac{M_{ud}}{M_d}\right)\left(\dfrac{f_{t\,adm}}{0.87f_y}\right)\left(1 + \dfrac{A_s'}{A_s}\right)\left(2\dfrac{d}{h} - 1\right)^2}.$$

If an estimate is made of the ratios z/h, A_s'/A_s, and d/h, this formula enables suitable leading dimensions b and h to be determined. The area of tensile and compressive reinforcement may then be calculated more accurately for the ultimate limit state, or obtained from the chart in Fig. 7.4.

EXAMPLE

Determine the minimum thickness of the vertical cantilever wall of a reinforced concrete water tank necessary to retain a depth of water of 4 m (13 ft) and calculate the required areas of tensile and compressive reinforcement. The characteristic yield stress of the reinforcing steel is 250 N/mm² (36 000 lbf/in²), the compressive strength of the concrete is 25 N/mm² (3600 lbf/in²) and at the serviceability limit state the tensile stress in the concrete must not exceed 1.85 N/mm² (270 lbf/in²).

Moment due to water pressure

$$M_d = \frac{1}{6} \times 1000 \times 4^3 \times 9.81 = 104\,500 \text{ Nm/m width (282 000 lbf in/ft width)}.$$

Using a factor $\gamma = 1.5$ for the ultimate limit state

$$M_{ud} = 1.5 \times 104\,500 = 157\,000 \text{ Nm/m width (425 000 lbf in/ft width).}$$

Assume $\alpha = 6\ z/h = 0.75\ A_s'/A_s = 0.5\ d/h = 0.9$.

$$\frac{M_d}{bh^2} \leqslant \frac{1.85}{6 - \dfrac{3(6-1)}{0.75}\,(1.5)\,\dfrac{(1.85)}{(0.87 \times 250)}\,(1+0.5)\,(2 \times 0.9 - 1)^2}$$

$$= 0.321 \text{ N/mm}^2 \text{ (47 lbf/in).}$$

$$\frac{104\,500 \times 1000}{1000 \times h^2} \leqslant 0.321,$$

$$h \geqslant 570 \text{ mm (22.4 in).}$$

Select an overall thickness of $h = 600$ mm (24 in) and 40 mm (1½ in) cover, so that $d \simeq 550$ mm and $d/h \simeq 0.90$.

$$\frac{M_{ud}}{f_{cu}\,bd^2} = \frac{157\,000 \times 1000}{25 \times 1000 \times 550^2} = 0.0208.$$

From Fig. 7.4,

$$\frac{f_y A_s}{f_{cu}\,bh} \simeq \frac{0.025 \times 25 \times 1000 \times 600}{250} = 1500 \text{ mm}^2\text{/m width (0.71 in}^2\text{/ft width).}$$

20 mm bars at 200 m centres provide an area of 1570 mm²/m width. (¾ in bars at 7½ provide 0.71 in²/ft width).

The reinforcement on the compressive face is not required for the ultimate limit state but is necessary for control of possible shrinkage cracking. Half the area of the tensile reinforcement has already been assumed; this may be provided by 10 mm bars at 100 mm centres (⅜ in bars at 4 in centres).

REFERENCES

[7.1] Magnel, G. (1948) *Prestressed Concrete*, Ch. 3, 19-26. London: Concrete Publications.
[7.2] Bennett, E. W. (November, 1959). A graphical method for the design of prestressed beams. *Concrete. Construct. Engineering*, **53**, 399-403.

8
Dimensioning for axial load and bending

Short axially loaded columns — ultimate limit state

Short columns are defined as those whose strength is unaffected by lateral instability. It has already been noted (Chapter 5, p. 140ff) that the theoretical criterion of lateral stability is the slenderness ratio of the effective length to the radius of gyration, but that in reinforced concrete the latter dimension is uncertain and variable.

The practical criterion commonly adopted is therefore the ratio of the effective length to the lateral dimension of the column in the direction of buckling; for example in the British Code CP 114 : 1957 short columns were defined as those for which the ratio of the effective length to the last lateral dimension is less than 15, but in the latest revision this has been reduced to 12.

On the basis of the present code assumptions, the design strength of an axially loaded column has been shown in Chapter 4, (p. 89) to be given by the formula

$$N_{ud} = 0.4 f_{cu} A_c + 0.67 f_y A_s{}'.$$

An arbitrary reduction of about 12% of the strength is made to accommodate the limited bending of a restrained column supporting an approximate symmetrical arrangement of uniformly loaded beams, so that

$$N_{ud} = 0.35 f_{cu} A_c + 0.6 f_y A_s{}'.$$

The minimum area of reinforcement is normally 1 per cent of the area of the section, but for lightly loaded columns it is given by

$$f_y A_s{}' \geq 0.15 N_{ud}.$$

The reinforcement ratio $A_s{}'/bh$ must not exceed 8% for horizontally cast columns, 6% for vertically cast columns and 10% at laps in the reinforcement; this makes allowance for the greater ease with which the concrete can be compacted in a horizontally cast member and for the local congestion caused by lapped bars.

The above design formulae are represented graphically on design charts in Figs. 8.1 and 8.2 and an example of their use is given below:

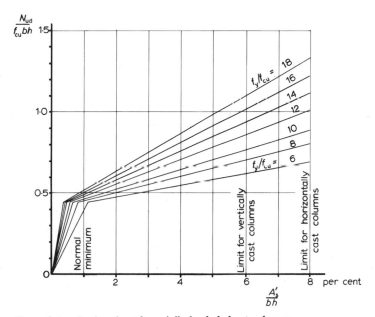

Figure 8.1   Design chart for axially loaded short columns

EXAMPLE

Design the central column for the reinforced concrete building frame in Fig. 8.3.

The characteristic load on the central column at the level of the top floor is made up as follows:

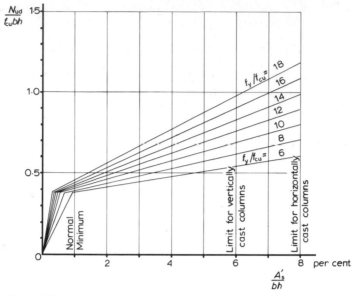

Figure 8.2   Design chart for columns supporting symmetrical beams

Self weight of column

(Assumed 300 mm x 300 mm): 0.3 x 0.3 x 4.0 x 2.4 x 9.81	=	8.5 kN.
R.C. beams: 0.3 x 0.45 x 7.2 x 2.4 x 9.81	=	22.9 kN.
Longitudinal R.C. beam: 0.3 x 0.35 x 3.7 x 2.4 x 9.81	=	9.2 kN.
Precast floor units: 4.0 x 7.2 x 300 x 9.81/1000	=	84.7 kN.
50 mm screed + waterproofing: 4.0 x 7.5 x 150 x 9.81/1000	=	44.1 kN.
Total dead load		169.4 kN.
Imposed load: 4.0 x 7.5 x 1.5	=	45.0 kN.
		214.4 kN
		(48 300 lbf).

At the level of the first and second floors:

Self weight of column (as before)	=	8.5 kN.
R.C. beam (,, ,,)	=	22.9 kN.
Longitudinal R.C. beam (,, ,,)	=	9.2 kN.
Pre-cast floor units (,, ,,)	=	84.7 kN.
Screed and floor surfacing: 4.0 x 7.5 x 140 x 9.81/1000	=	41.1 kN.
Allowance for partitions, etc. 4.0 x 7.5 x 2.0	=	60.0 kN
Total dead load		226.4 kN.
Imposed load: 4.0 x 7.5 x 3.0	=	90.0 kN.
		316.4 kN
		(71 200 lbf).

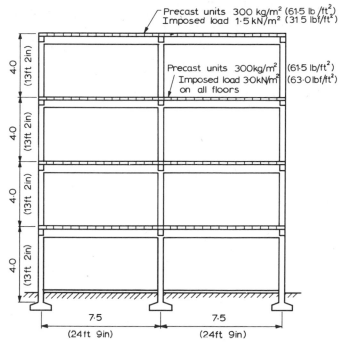

Figure 8.3   Example — Loading on reinforced concrete column

At the base of the column the dead load will be slightly increased by the greater storey height and will be 227.4 kN. Using the prescribed load factors of 1.4 and 1.6 the design ultimate column load at each level will be as follows:

3rd floor:

$$169.4 \times 1.4 + 45.0 \times 1.6 \ = \ 309.2 \text{ kN.}$$
$$226.4 \times 1.4 + 90.0 \times 1.6 \ = \ \underline{461.0 \text{ kN.}}$$
$$770.2 \text{ kN.}$$

2nd floor:

$$226.4 \times 1.4 + 90.0 \times 1.6 \ = \ \underline{461.0 \text{ kN}}$$
$$1231.2 \text{ kN}$$

1st floor:

$$227.4 \times 1.4 + 90.0 \times 1.6 \ = \ \underline{462.4 \text{ kN}}$$

Base

$$\underline{1693.6 \text{ kN}}$$
$$(382\ 000 \text{ lbf})$$

The central column in this frame may be regarded as restrained and supporting a symmetrical arrangement of beams. Between the base and first floor the effective height is about 0.7 of the height between the restraints;

therefore if the ratio of effective height to least lateral dimension is not to exceed 12, the minimum width of columns will be:

$$\frac{4500 \times 0.7}{12} = 263 \text{ mm} \qquad 10.4 \text{ in.}$$

A column of cross section 300 mm (12 in) square is therefore satisfactory.

Specifying concrete of characteristic strength 30 kN/mm² (4350 lbf/in²), resistance of columns without reinforcement

$$= 0.35 \times 30 \times 300 \times 300 \times 10^{-3} = 945 \text{ kN (212 500 lbf)},$$

additional resistance to be provided by reinforcement

$$= 1694 - 945 = 749 \text{ kN (168 500 lbf).}$$

The stress developed in steel reinforcement of characteristic strength 410 N/mm² (60 000 lbf/in²) is 0.6 × 410 = 246 N/mm² (36 000 lbf/in²). Since, however, the reinforcement displaces an equal area of concrete, the differential stress in the reinforcement must be used in calculating the area required to provide the above additional resistance.

Differential stress in reinforcement

$$= 246 - 0.35 \times 30 = 235.5 \text{ N/mm}^2 \text{ (34 200 lbf/in}^2\text{).}$$

Area of reinforcement required

$$= \frac{749 \times 10^3}{235.5} = 3180 \text{ mm}^2 \text{ (4.95 in}^2\text{).}$$

8–25 mm bars provide 3930 mm² (8–1 in bars provide 6.3 in²). Alternatively the design chart, Fig. 8.2 may be used, and

$$\frac{N_{ud}}{f_{cu} bh} = \frac{1694 \times 10^3}{30 \times 300 \times 300} = 0.628,$$

$$\frac{f_y}{f_{cu}} = \frac{410}{30} = 13.7.$$

Hence from the chart $A_s'/bh = 3.6$ per cent,

$$A_s' = 300 \times 300 \times 0.036 = 3240 \text{ mm}^2.$$

Although the overall amount of computation is not greatly reduced, the chart is useful because it indicates both the maximum load capacity of the column and the load attainable with minimum reinforcement. It is a useful tool for the dimensioning of a cross section which is not governed by slenderness.

The size and spacing of the transverse reinforcement are determined by general rules. Using the British Code, the size of transverse reinforcement must be at least one-quarter the size of the largest compression bar $= \frac{1}{4} \times 25 = 6.2$ mm.

Applying the rule strictly, 8 mm binders are required, although it would appear rather arbitrary to reject the use of 6 mm binders with 25 mm main bars, as these sizes correspond closely to the common use of $\frac{1}{4}$ in binders with 1 in main bars in the Imperial sizes.

The maximum spacing of the transverse reinforcement is 12 times the size of the smallest compression bar $= 12 \times 25 = 300$ mm.

Two possible arrangements of links are shown in Fig. 8.4. In order to fulfil their function of restraining the main reinforcement against transverse movement in any direction, the binders are arranged to support each bar in an angle. This angle should not be greater than 135° (except for circular binders), and although this corner support need not now be provided for alternate bars on the face of a column, such bars should not be more than 150 mm (6 in) from a bar provided with corner support.

Obviously the amount of main reinforcement may be reduced in the less heavily loaded columns supporting the upper floors and roof, and a smaller cross

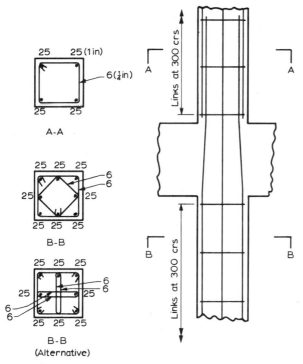

Figure 8.4   Example – Details of R.C. column and joint at first floor

section could also be used, although in a small building, such as the one in the present example, it would be more economical to have the same size of column throughout to enable the same shuttering to be used at each storey. Similarly practical economic considerations should be kept in mind when reducing the reinforcement in the upper storeys. In our example the reinforcement just above first floor level may be met by four 25 mm bars which is a convenient arrangement obtained by simply discontinuing the four bars at the centre of the faces. However, although in the storey above, the reinforcement could be reduced to four bars of the minimum allowable diameter of 12 mm, this would be a very doubtful economy since the spacing of the binders would have to be reduced to $12 \times 12 = 144$ mm, resulting in an increased cost of material and labour.

It is usual to form a lapped joint of the main column reinforcement at each floor level to avoid the inconvenience of long bars projecting vertically above the work during construction. A typical example of such a joint is shown in Fig. 4 from which it will be seen how the ends of the main reinforcement from the lower column are left projecting as 'starter bars' for the column above. The length of lap is calculated from the anchorage bond stress as explained in Chapter 2 (p. 54ff).

Members in axial tension

Tie members loaded in axial tension are not often designed in reinforced concrete, since in the absence of compression the concrete is unable to fulfil a structural function. This limitation does not apply to prestressed concrete, however, and there have been some notable examples of the use of prestressing for trusses, although this type of prestressed structure has not been widely adopted. An example in which direct tension is unavoidable is in the circumferential loading of cylindrical tanks, silos and pipes; here both reinforced and prestressed concrete are employed and the latter offers particular advantages.

The ultimate load for the limit state is readily obtained since the concrete section must be considered to be completely severed by a crack at the critical section so that the entire axial load is resisted by the reinforcement and/or tendons; these will normally be arranged symmetrically and therefore will be uniformly stressed. The total area of steel A_s is therefore determined by this limit state.

$$A_s = \frac{N_{ud}}{0.87 f_y} \text{ for reinforced concrete,}$$

$$A_p = \frac{N_{ud}}{0.87 f_{pu}} \text{ for prestressed concrete,}$$

where

N_{ud} = design tensile load at ultimate limit state,

f_y = yield stress of reinforcement,

f_{pu} = tensile strength of steel in tendons.

The serviceability limit state will generally be governed by the crack width, and the formulae given for flexural members are probably valid, although little experimental work appears to have been done on the width of cracks in reinforced concrete members subject to axial tension. In the parts of a liquid retaining structure which are in contact with the liquid, the serviceability limit state is usually taken to be the point at which cracks first appear. In the British Code the direct tensile strength of the concrete is assumed to be about $0.39\sqrt{f_{cu}}$ N/mm^2 where f_{cu} is the cube strength in N/mm^2, so that with a factor of 1.3 the design stress at the serviceability limit state would be $0.30\sqrt{f_{cu}}$.

In prestressed concrete the tensile stress criterion may also be used for the serviceability limit state and the allowable stress would be zero for Class 1 members and $0.30\sqrt{f_{cu}}$ (as above) for Class 2, but with the present state of knowledge it would appear imprudent to design Class 3 members for direct tension.

It must be pointed out, however, that the use of a constant allowable tensile stress in design does not give a constant load factor against cracking. This may be demonstrated as follows:

Let

N_r = tensile load which would cause cracking,

N_d = design load = characteristic load × 1.0,

A = cross-sectional area of concrete section,

f_p = compressive prestress in concrete,

f_{ct} = tensile strength of concrete.

$$\frac{N_r}{A} = f_p + f_{ct}.$$

Using an allowable stress of $f_{ct}/1.3$ as in the British Code

$$\frac{N_d}{A} = f_p + \frac{f_{ct}}{1.3}.$$

Now the true load factor against cracking, γ_r is

$$\gamma_r = \frac{N_r}{N_d} = \frac{f_p + f_{ct}}{f_p - f_{ct}/1.3}.$$

It will be seen from this expression that the load factor is 1.3 where there is no prestress but where the prestress is finite, the load factor decreases with the value of the prestress and is only a little greater than unity for the normal level of compressive prestress (Fig. 8.5). A more soundly based procedure is that recommended in the B.S. Code for liquid-returning structures (CP 2007 : 1960) where it is specified that the 'load factor against cracking', defined as the ratio of the total (dead + live) load at cracking to the total (dead + live) working load, should be not less than 1.25. On this basis the allowable tensile stress $f_{t\,adm}$ in the concrete is given by the expression

$$f_{t\,adm} = \frac{f_{ct} - (\gamma_r - 1)f_p}{\gamma_r},$$

where γ_r is the load factor against cracking, which in this Code is specified as 1.25.

A formula giving the required area of concrete for a prestressed section may be derived from the stress conditions by a parallel method to that used in the previous chapter to obtain the formula for the required section modulus.

Compressive stress under minimum tensile load

$$f_c - \frac{N_{min}}{A} \leqslant f_{c\,p\,adm}.$$

Tensile stress under maximum tensile load

$$-\eta f_c + \frac{N_d}{A} \leqslant f_{t\,adm}.$$

Combining the two stress conditions

$$A \geqslant \frac{N_d - \eta N_{min}}{\eta f_{cp\,adm} + f_{t\,adm}},$$

where

A = equivalent area of concrete section,

N_d = design tensile load,

N_{min} = minimum (sustained) tensile load, frequently zero,

f_c = compressive prestress in concrete,

$f_{cp\,adm}$ = permissible compressive stress in concrete at transfer prestress,

$f_{t\,adm}$ = permissible tensile stress in concrete under working load,

η = prestress loss ratio.

The formula may be adapted for the limit state of tensile cracking by replacing N_r by the design cracking load and $f_{t\,adm}$ by the tensile strength of the concrete, as in the following example.

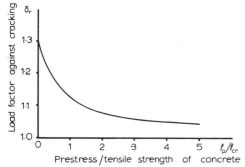

Figure 8.5 Load factor against cracking with constant allowable tensile stress of $f_{ct}/1.3$

EXAMPLE

Calculate the thickness and circumferential reinforcement of the concrete wall of a cylindrical tank of diameter 10 m (32.8 ft) with a frictionless sliding joint at the base, to withstand a 5 m (16.4 ft) head of water.

Pressure at base of wall

$$= 1000 \times 5 \times 9.81 \text{ N/m}^2 = 48 \text{ kN/m}^2 \text{ (1000 lbf/ft}^2).$$

Circumferential tensile force

$$= 48 \times \frac{10}{2} = 240 \text{ kN/m (16 400 lbf/ft)}.$$

(a) *Design in reinforced concrete*

Allow a load factor of 2.5 against ultimate failure and 1.25 against cracking. Specify concrete of characteristic strength 40 N/mm² (5800 lbf/in²) and mild steel reinforcement of characteristic yield stress 250 N/mm² (36 400 lbf/in²). Mild steel is used in preference to high strength steel to provide better control of cracking through the larger area required.

Area of steel required for ultimate limit state

$$= \frac{240 \times 1000 \times 2.5}{0.87 \times 250} = 2760 \text{ mm}^2/\text{m} \ (1.30 \text{ in}^2/\text{ft}).$$

16 mm bars at 140 mm centres on each face give 2860 mm²/m ($\frac{5}{8}$ in bars at 5 in centres give 1.47 in²/ft).

Design tensile strength of concrete $= 0.30\sqrt{40} = 1.90 \text{ N/mm}^2 \ (275 \text{ lbf/in}^2)$

Equivalent area of concrete required for serviceability limit state

$$= \frac{240 \times 1000 \times 1.25}{1.90} = 158 \ 000 \text{ mm}^2/\text{m} \ (75 \text{ in}^2/\text{ft}).$$

Equivalent increase of area of concrete due to steel ($\alpha = 15$)

$$= 2860 \times (15 - 1) = 40 \ 000 \text{ mm}^2/\text{m} \ (18.9 \text{ in}^2/\text{ft}).$$

Required nominal area of section

$$= 158\,000 - 40\,000 = 118\,000 \text{ mm}^2/\text{m} \ (56.1 \text{ in}^2/\text{ft}).$$

Required thickness of wall

$$= 118 \text{ mm, say } 125 \text{ mm (5 in).}$$

(b) Design in prestressed concrete

The area of the required concrete section is first calculated. Allowing a compressive prestress of 16 N/mm² (2300 lbf/in²) at the time of prestressing and 15% loss of prestress

$$A \geqslant \frac{N_d - \eta N_{min}}{\eta f_{cp\ adm} + f_{t\ adm}},$$

therefore

$$A \geqslant \frac{240\,000 \times 1.25 - 0}{0.85 \times 16 + 1.90} = 19\,300 \text{ mm}^2/\text{m} \ (9.1 \text{ in}^2/\text{ft}).$$

This is equivalent to a wall less than 20 mm thick, demonstrating the great advantage of prestressing in concrete sections designed to withstand direct tension.

A minimum thickness is dictated by practical considerations, such as the ability to place and compact high strength concrete and the need for water-tightness. For these reasons a wall thickness of 100 mm will be adopted.

Prestress required

$$f_c \geqslant \frac{1}{\eta} \left(\frac{N_d}{A} - f_{t\ adm} \right),$$

therefore

$$f_c \geqslant \frac{1}{0.85} \left(\frac{240\,000 \times 1.25}{100 \times 1000} - 1.90 \right) = 1.29 \text{ N/mm}^2 \ (185 \text{ lbf/in}^2).$$

A minimum prestress of 3.5 N/mm² (510 lbf/in²) is adopted.
Prestressing force = 100 × 1000 × 3.5 = 350 000 N/m (23 9000 lbf/ft).

In the case of prestressing by means of a continuous winding of 5 mm hard drawn wire of tensile strength 1500 N/mm² (215 000 lbf/in²) at a stress of 950 N/mm² (130 000 lbf/in²) immediately after winding.

Number of wires per metre

$$= \frac{350\,000}{19.6 \times 950} = 18.8.$$

Pitch of winding

$$= \frac{1000}{18.8} = 53 \text{ mm (2.1 in).}$$

Total tensile force required for ultimate limit state

$$= 240 \times 2.5 = 600 \text{ kN/m (41 000 lbf/ft).}$$

Tensile strength of prestressed steel

$$= \frac{1000}{53} \times 19.6 \times \frac{1500}{1000} = 555 \text{ kN/m} \quad (37\,900 \text{ lbf/ft).}$$

Additional tensile resistance required

$$= 600 - 555 = 45 \text{ kN/m (3080 lbf/ft).}$$

Area of additional reinforcement

$$= \frac{45 \times 1000}{0.87 \times 250} = 207 \text{ mm}^2/\text{m (0.098 in}^2/\text{ft),}$$

6 mm bars on each face at 250 mm centres provide 225 mm^2/m (¼ in bars at 10 in centres provide 0.188 in^2/ft).

Combined axial force and flexure — ultimate limit state

The dimensioning of members subject to any combination of axial force, either compressive or tensile, and to bending, can be based on a single set of assumptions founded on the experimentally observed behaviour discussed in Chapter 4. The assumptions here used are substantially those of the British Code, which are as follows:

1. The strain distribution in the concrete is derived from the assumption that plane sections remain plane.
2. The stresses in the concrete in compression are either derived from the stress-strain curve given in Fig. 4.2 with $\gamma_m = 1.5$ or taken as equal to 0.4 f_{cu} over the whole compression zone; in both cases the strain at the outermost fibre is taken as 0.0035.
3. The tensile strength of the concrete is ignored.
4. The strain in the reinforcement, whether in compression or tension, is derived from the assumption that plane sections remain plane.
5. The stresses in the reinforcement are derived from the stress-strain curves given in Fig. 4.2 with $\gamma_m = 1.15$.

It will be shown that at certain points these assumptions require some qualification. In calculations the second alternative stress distribution in assumption 2 will be adopted for ease of working; a calculation indicates that the maximum difference in the moment of the concrete in compression derived from this, as opposed to the parabolic-rectangular stress distribution, is about 11% on the safe side, but that it is usually much less.

Typical stress and strain distribution diagrams based on the above assumptions for all combinations of load and moment from axial compression to axial

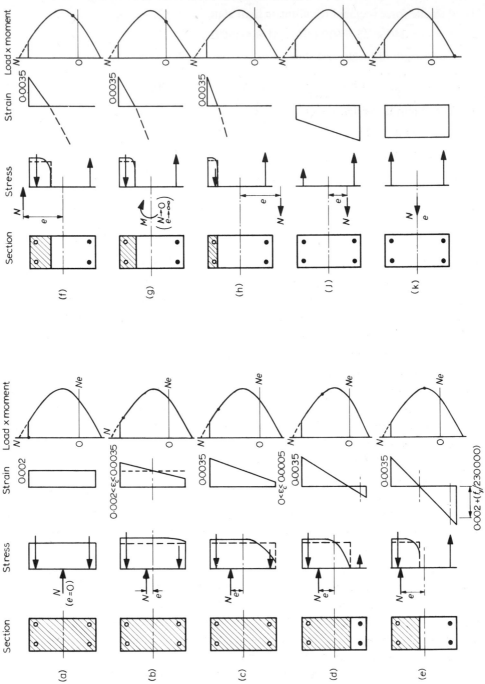

Figure 8.6 Stress and strain distribution in members subject to combined axial force and bending

tension are shown in Fig. 8.6. The position of the point on the load-moment diagram corresponding to each combination is also shown. The various cases will be briefly reviewed before going on to consider the dimensioning of sections, and reinforcement.

Figure 8.6(a) illustrates the case of axial compression, already considered. The ultimate compressive strain of 0.0035 recommended in the British Code is too large either for this condition or the succeeding one (Fig. 8.6(b)) for a slightly eccentric compressive load. The strain assumption used here is therefore that of the European Concrete Committee, whereby the strain is assumed to have a mean value (i.e. the value at the centroid of the section) of 0.0020 until the maximum compressive strain attains the value of 0.0035. With increasing eccentricity of the compressive force the maximum compressive strain at the opposite face decreases (Fig. 8.6(c)). Throughout these three stages the compressive stress in the concrete is either obtained from the strain using the idealised stress-strain diagram of Fig. 4.2 or is assumed to have a design value of $0.4 f_{cu}$ uniformly distributed across the whole section, while the design stress in the steel is related to the assumed strain by the stress-strain characteristic.

As the eccentricity e of the load N becomes still greater, the minimum strain becomes tensile (Fig. 8.6(d)), and shortly afterwards the stress in the steel nearest to this face will also become tensile.

Over the next stage the compressive zone of the concrete extends only to the depth of the neutral axis, any tensile stress in the concrete being neglected. The stress in the steel is normally at the maximum design value on the compressive side and at a value less than the yield point, obtained from the strain distribution and stress-strain characteristic, at the tensile side. The section is thus 'over-reinforced'.

The 'balanced' condition (Fig. 8.6(e)) is attained when the tensile steel just begins to yield. According to the assumed tensile stress-strain characteristic (Fig. 4.2) the design value of the strain in the steel will be $0.002 + f_y/230\,000$ and since the strain in the concrete at the compression face is assumed to be 0.0035, the position of the neutral axis is defined and is the same for all percentages of reinforcement, although the magnitude of the force N and eccentricity e will be related to the amount of compressive and tensile reinforcement.

Beyond the balanced point the section is 'under-reinforced' (Fig. 8.6(f)). The tensile reinforcement remains at the design yield stress ($0.87 f_y$) but the position of the neutral axis can no longer be determined on the basis of a linear strain distribution ('plane sections remain plane'). This is firstly because the steel strain at yield is indeterminate, on account of the horizontal plateau in the stress-strain characteristic. In addition, as discussed in Chapter 4, the effect of bond on the steel in the region of a tensile crack is such that in an under-reinforced section the steel stress may be considerably greater than the value corresponding to the average linear strain. In calculations the position of the neutral axis must therefore be established from the conditions of equilibrium.

The case of pure moment, which has been discussed at length in Chapter 4, and the dimensioning of sections, considered in Chapter 7, may be regarded as a limiting case of eccentric load in which $N \to 0$ and $e \to \pm\infty$. In Fig. 8.6(g) the pure moment case is shown for an under-reinforced section but it may of course be over-reinforced when the area of tensile reinforcement is large.

The last three diagrams in the series illustrate the effect when the force N is tensile. When the eccentricity is very large, as in Fig. 8.6(h) there may still be a shallow zone of compression in the concrete, but this vanishes as the load becomes more nearly axial (Fig. 8.6(j)). The stress in the reinforcement on the side of maximum tension will be the design yield value and the reinforcement on the other side will be at a lower tensile stress. Although a linear tensile strain distribution is apparent on the diagram it will be shown that this assumption is not required for calculating reinforcement.

The final diagram, Fig. 8.6(k), is the case of an axial tensile load which has already been discussed.

Preliminary dimensioning of sections subject to compression and bending.

A useful tool for the preliminary dimensioning of members in the region between axial compression and pure flexure is the curve representing the relationship between $N/f_{cu}bh$ and e/h. It will be found that if the dimensions are such as to lie within a limiting curve represented by the equation

$$\frac{N}{f_{cu}\,bh} = 0.5 - 0.55 \left(\frac{e}{h}\right) + 0.5 \left(\frac{e}{h}\right)^2$$

shown in Fig. 8.7, then the total amount of compressive and tensile reinforcement will usually be about 2–3%.

Moreover, the parameter $N/f_{cu}be$ is represented by a series of straight lines, so that if one is designing for given values of N and e and has specified the characteristic cube strength f_{cu}, it is possible to read off from the chart a suitable value of h for any required value of b, or vice-versa, as will be shown in an example below.

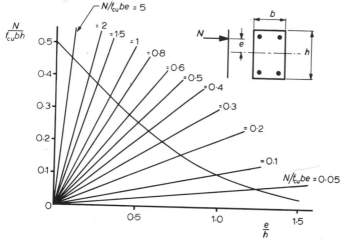

Figure 8.7   Chart for preliminary dimensioning of rectangular sections for compression and bending

Dimensioning unsymmetrical reinforcement at sections subject to combined axial load and flexure.

When the dimensions of the section have been determined the required amounts of reinforcement for compression and bending may be found by means of design charts in the form of load-moment curves for varying reinforcement percentages and cover ratios. A number of charts is required, especially if sections of different shapes and unsymmetrical arrangements of reinforcement are to be included. However, it is not difficult to calculate the reinforcement directly when the general principles and behaviour discussed above are understood, and this procedure is far more instructive for the student.

Before giving examples of calculations the basic steps will be described. These are illustrated by Fig. 8.8 and it will be noted that the cover over the reinforcement must be assumed at the commencement.

Step 1.

The depth of the compressive zone x is calculated from the condition of equilibrium of moments about the level of the tensile reinforcement.

Referring to Fig. 8.8, the moment of the external force is $N(e + d - h/2)$. The maximum opposing moment of the concrete is $0.4 f_{cu} bh (d - h/2)$ when the whole section is in compression (Fig. 8.8(a)). Therefore if

$$N \left(e + d - \frac{h}{2} \right) > 0.4 f_{cu} bh \left(d - \frac{h}{2} \right)$$

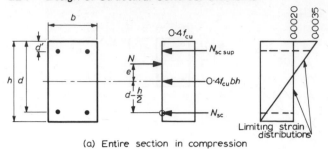

(a) Entire section in compression

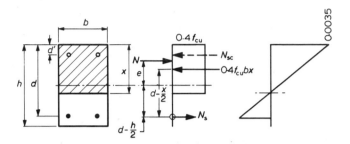

(b) Part of section in tension

Figure 8.8   Dimensioning of reinforcement for rectangular reinforced concrete section subject to axial load and bending

the condition may be that shown in Fig. 8.6(b) or (c) and compressive reinforcement must be added to balance the external moment. The possibility of this solution depends on the strain distribution, which will be checked later.

If, on the other hand, the external moment can be balanced with only part of the section in compression, it will be seen from Fig. 8.8(b) that the equilibrium equation is

$$N \left(e + d - \frac{h}{2} \right) = 0.4 \, bx \left(d - \frac{x}{2} \right).$$

The depth of compressive zone required is obtained from the solution of this quadratic equation, which is the same as that for pure bending given in the previous chapter (p. 178),

$$\frac{x}{d} = 1 - \sqrt{1 - \frac{5 \, M}{f_{cu} \, bd^2}} \quad ,$$

where in this case

$$M = N \left(e + d - \frac{h}{2} \right).$$

The above calculations are for a rectangular section, but the principles may readily be applied to any shape of section. The computations for some shapes will be more involved and may entail a procedure of trial and error.

Step 2.

The required resistance of the compressive reinforcement is calculated, also from the condition of equilibrium of moments.

Compressive reinforcement will be necessary when the external moment cannot be balanced by the compressive resistance of the whole concrete section. From the moment equilibrium condition (Fig. 8.8(a) this force is given by

$$N_{\text{sc sup}} = \frac{N(e + d - h/2) - 0.4\,f_{\text{cu}}\,bh(d - h/2)}{d - d'}.$$

If, after checking the strain in Step 4, it is found that the lower part of the section must be in tension, this step has to be reworked with an assumed value of x (Fig. 8.8(b)). Then

$$N_{\text{sc sup}} = \frac{N(e + d - h/2) - 0.4\,f_{\text{cu}}\,bx\,(d - x/2)}{d - d'}.$$

Step 3.

The force to be provided by the bottom reinforcement is calculated from the condition that the forces normal to the section must be in equilibrium.

When the whole section is in compression (Fig. 8.8(a)), the value of the required compressive force is

$$N_{\text{sc}} = N - 0.4\,f_{\text{cu}}bh - N_{\text{sc sup}}.$$

A negative result will denote a tensile force, and this case is discussed below.

When only a part of the section is in compression a tensile force will be required (Fig. 8.8(b)) the value of which is

$$N_{\text{s}} = 0.4\,f_{\text{cu}}bx + N_{\text{sc sup}} - N.$$

The second term will be omitted when the compressive reinforcement is absent or only nominal.

Step 4.

The stress in the reinforcement is calculated from the strain distribution, adjusting the value of x and reworking Steps 2 and 3 if necessary. The required areas of reinforcement may then be obtained.

Top reinforcement.
The stress in the top compressive reinforcement is related to the strain by the idealised stress-strain characteristic given in Fig. 4.2, but since the strain at the level of this reinforcement is almost always greater than 0.002, the maximum design compressive stress of $\{2000/(2300 + f_y)\}f_y$ may be assumed.

The reduction factor $2000/(2300 + f_y)$ for the compressive yield stress of the steel seems unjustifiably elaborate and a constant factor of 0.72 is normally adequate. Also, to be strictly consistent with the methods of calculation for axial compression the differential stress in the compressive reinforcement, $\{2000/(2300 + f_y)\}f_y - 0.4 f_{cu}$ should be used to allow for the area of concrete displaced by the steel, although this is not done in the British Code formulae for pure bending with compressive reinforcement.

Bottom reinforcement.
When the whole of the concrete section is in compression the strain in the bottom reinforcement can assume any compressive value up to 0.002, as may be necessary to maintain equilibrium (Fig. 8.8(a)). The latter value, with the steel stress at its maximum design value of $0.87 f_y$ will yield the most economical value of A_s.

If, however, having initially assumed the whole section to be in compression, it is then found that the bottom reinforcement must be in tension to maintain equilibrium, steps 2 and 3 must be reworked with a value of x less than the depth of the section. A suitable trial value might be $x = 0.75d$ for which the design stress in the steel would be about 230 N/mm^2. If the area of tensile reinforcement calculated for this position of the neutral axis is uneconomically large, a further calculation may be made for the balanced condition for which $x = 805/(1265 + f_y)$ and the design stress in the steel is the maximum value of $0.87 f_y$.

When a value of x less than d has been calculated in Step 1 the tensile stress is checked from the strain diagram and the area of reinforcement calculated in the same way (Fig. 8.8(b)).

EXAMPLE: ECCENTRICALLY LOADED COLUMN

(a) Calculate suitable section dimensions and reinforcement for a column required to support a design ultimate load of 3500 kN (787 000) lbf) acting at an eccentricity of 40 mm (1.57 in), and

(b) determine the reinforcement for the same size of column to support a load of 2400 kN (540 000 lbf) at an eccentricity of 100 mm (3.93 in).

(a) The specified characteristic cube strength of the concrete is 37.5 N/mm^2 (5450 lbf/in^2) and the characteristic tensile strength of the steel 410 N/mm^2 ($59\ 500 \text{ lbf/in}^2$).

From the preliminary dimensioning formula (p. 222)

$$\frac{3\,500\,000}{37.5\,h^2} = 0.5 - 0.55 \times \frac{40}{h} + \frac{0.15 \times 1600}{h^2}.$$

$$h^2 - 44\,h - 186\,500 = x,$$

$$h \simeq 433 \text{ mm (17.0 in)}.$$

Select a 400 mm (15.7 in) square column and allow for 32 mm (1¼ in bars) with 25 mm (1 in) cover (Fig. 8.8(a)).

Moment of external load about level of bottom reinforcement

$$= 3500 \times 199/1000 = 696 \text{ kN m (6 150 000 lbf in)}.$$

Compressive resistance of full concrete section

$$= 0.4 \times 37.5 \times 400 \times 400/1000 = 2400 \text{ kN (540 000 lbf)}.$$

Moment of resistance of concrete about bottom reinforcement

$$= 2400 \times 159/1000 = 382 \text{ kN m (3 380 000 lbf in)}.$$

Required moment of resistance of top reinforcement

$$= 696 - 382 = 314 \text{ kN m (2 770 000 lbf in)}.$$

Required compressive resistance of top reinforcement

$$= \frac{314 \times 1000}{318} = 998 \text{ kN (222 000 lbf)}.$$

Required compressive force in bottom reinforcement

$$= 3500 - 2400 - 988 = 112 \text{ kN (2500 lbf)}.$$

Compressive stress in reinforcement

$$= \left(\frac{2000}{2300 + 410}\right) \times 410 = 302 \text{ N/mm}^2 \text{ (43 800 lbf/in}^2\text{)}.$$

Required area of top reinforcement

$$= \frac{988 \times 1000}{302} = 3260 \text{ mm}^2 \text{ (5.05 in}^2\text{)}.$$

Select 4–32 mm bars giving an area of 3200 mm² (4–1¼ in bars give 4.90 in²).
Required area of bottom reinforcement

$$= \frac{112 \times 1000}{302} = 371 \text{ mm}^2 \text{ (0.57 in}^2\text{)}.$$

Select 2—20 mm bars giving an area of 628 mm² (2—¾ in bars give 0.88 in²).

(b) Fig. 8.9 (b)

Moment of external load about level of bottom reinforcement

$$= 2400 \times 259/1000 = 622 \text{ kN m } (5\ 500\ 000 \text{ lbf in}).$$

Compressive resistance of concrete (as above)

$$= 2400 \text{ kN } (540\ 000 \text{ lbf}).$$

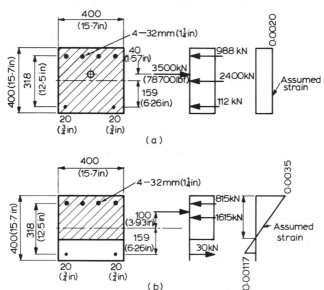

Figure 8.9   Example – Eccentrically loaded column

It is clear from the equilibrium condition of the normal forces that the bottom steel must be in tension.

Try

$$x = 0.75d$$

$$= 0.75 \times 359 = 269 \text{ mm } (10.6 \text{ in}).$$

Compressive resistance of concrete

$$= 0.4 \times 37.5 \times 400 \times 269/1000 = 1615 \text{ kN } (364\ 000 \text{ lbf}).$$

Moment of resistance of concrete

$$= 1615\ (359 - 269/2)/1000 = 363 \text{ kN m } (3\ 210\ 000 \text{ lbf in}).$$

Required compressive resistance of top reinforcement

$$= \frac{(622 - 363)}{318} \times 1000 = 815 \text{ kN } (183\ 000 \text{ lbf}).$$

Required tensile resistance of bottom reinforcement

$$= 1615 + 815 - 2400 = 30 \text{ kN } (6700 \text{ lbf}).$$

Required area of top reinforcement

$$= \frac{815 \times 1000}{302} = 2700 \text{ mm}^2 \ (4.19 \text{ in}^2).$$

Select 4—32 mm bars as above.

Tensile strain in bottom reinforcement

$$= \frac{0.0035(1 - 0.75)}{0.75} = 0.00117.$$

Tensile stress in bottom reinforcement (see Fig. 4.2)

$$= 200 \times 0.00117 \times 1000 = 234 \text{ N/mm}^2 \ (34\ 000 \text{ lbf/in}^2).$$

Required area of bottom reinforcement

$$= \frac{30 \times 1000}{234} = 128 \text{ mm}^2 \ (0.20 \text{ in}^2).$$

Select 2—10 mm bars or 2—20 mm bars as before giving an area of 156 mm²
(2—⅜ in bars give 0.22 in²).

EXAMPLE: PORTAL FRAME MEMBER.

*Design the cross section and reinforcement for a member 200 mm (7.9 in) wide
in a portal frame, subject to a moment of 125 kN m (1 105 000 lbf in) and a
thrust of 300 kN (67 500 lbf).*

Eccentricity

$$e = \frac{125 \times 1000}{300} = 417 \text{ mm } (16.4 \text{ in}).$$

Using steel and concrete of the same strength as in the previous example

$$\frac{N}{f_{cu} \, be} = \frac{300 \times 1000}{37.5 \times 200 \times 417} = 0.096.$$

The preliminary dimensioning chart (Fig. 8.7) indicates $e/h \simeq 1.0$. Let depth
$h = 450$ mm (17.7 in) and assume 25 mm (1 in) bars with 25 mm (1 in) cover
(Fig. 8.10).

Moment of thrust about bottom reinforcement

$$= 300 \ (417 + 187)/1000 = 181 \text{ kN m } (1\ 600\ 000 \text{ lbf/in}).$$

Depth of concrete in compression

$$x = \left[1 - \sqrt{\left(1 - \frac{5 \times 181\,000\,000}{37.5 \times 200 \times 412} \right)} \right] = 190 \text{ mm (7.5 in)}.$$

Compressive resistance of concrete

$$= 0.4 \times 37.5 \times 200 \times 190/1000 = 570 \text{ kN (105 000 lbf)}.$$

No compressive top reinforcement is required.

Required tensile resistance of bottom reinforcement

$$= 570 - 300 = 270 \text{ kN (60 700 lbf)}.$$

Stress in steel = maximum design value.

Required area of bottom reinforcement

$$= \frac{270 \times 1000}{0.87 \times 410} = 758 \text{ mm}^2 \ (1.17 \text{ in}^2).$$

Use 2-25 mm bars giving an area of 982 mm² (2-1 in bars give 1.57 in²).

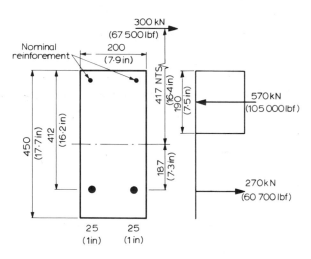

Figure 8.10   Example — Portal frame member

EXAMPLE: TIE MEMBER IN REINFORCED CONCRETE TRUSS

Calculate the reinforcement required for a tie member of cross section 200 × 200 mm (7.9 × 7.9 in) in a reinforced concrete truss to withstand an ultimate tensile force of 150 kN (33 800 lbf) at an eccentricity of 40 mm (1.6 in).

Allow for 15 mm (⅝ in) cover over 20 mm (¾ in) bars (Fig. 8.11).
Moment of external force about level of bottom reinforcement

$$= 150 \times 35 = 5250 \text{ kN m } (46\,500\,000 \text{ lbf in}).$$

Tensile resistance of top reinforcement

$$= \frac{5250}{150} = 35 \text{ kN } (7900 \text{ lbf}).$$

Tensile resistance of bottom reinforcement

$$= 150 - 35 = 115 \text{ kN } (25\,900 \text{ lbf}).$$

Tensile stress in reinforcement

$$= 0.87 \times \text{yield stress}$$
$$= 0.87 \times 410 = 357 \text{ N/mm}^2 \ (52\,000 \text{ lbf/in}^2).$$

Required area of top reinforcement

$$= \frac{35 \times 1000}{357} = 98 \text{ mm}^2 \ (0.13 \text{ in}^2).$$

Use 2-8 mm bars giving an area of 100 mm² (2-⅜ in bars give 0.22 in²).
Required area of bottom reinforcement

$$= \frac{115 \times 1000}{357} = 322 \text{ mm}^2 \ (0.50 \text{ in}^2).$$

Use 2-16 mm bars giving an area of 402 mm² (2-⅝ in bars give 0.61 in²).

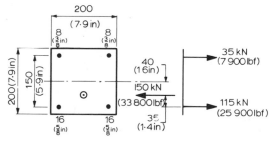

Figure 8.11   Example – Tie member in reinforced concrete truss

Dimensioning of symmetrical reinforcement at sections subject to combined axial load and flexure

Reinforced concrete sections subject to compression and bending are commonly
reinforced by bars arranged symmetrically with an equal area of steel on each
side of the axis of bending. This arrangement is particularly advantageous when

the eccentricity of the load is uncertain or variable; for example one would hesitate to use such a small percentage of steel as that calculated for the bottom steel in Fig. 8.9 if there was any risk of a more unfavourable combination of load and moment.

Most load-moment design charts are compiled for symmetrically reinforced sections, but a good approximation of the required area of symmetrical reinforcement may be obtained by direct calculation, with a slight modification of the method used for unsymmetrical reinforcement.

When the whole of the concrete section is in compression the area of top reinforcement may be found from the condition of moment equilibrium, as described for the calculation of unsymmetrical reinforcement (Steps 1 and 2, p. 223ff). An equal area of bottom reinforcement is provided, which will only be lightly stressed if the load is eccentric.

When only part of the section is in compression, Step 1 must be modified, and the depth of the compressive zone (x) is found from the condition of equilibrium of the forces normal to the section, making use of the fact that, when the reinforcement is symmetrical and x is approximately $0.5\,d$, the compressive force in the top reinforcement is approximately balanced by the tensile force in the bottom reinforcement. Thus

$$N \simeq 0.4\,f_{cu}\,bx,$$

$$x \simeq \frac{N}{0.4\,f_{cu}\,b}\,.$$

Step 2 is unchanged, the required compressive resistance of the top reinforcement ($N_{sc\,sup}$) being found, as previously, from the condition of equilibrium of moments.

Step 3 is no longer applicable, but if x is greater than about $0.5\,d$ the stress in the bottom reinforcement will have to be checked as in Step 4 previously.

EXAMPLES

Calculate the areas of steel required in the three previous examples if the reinforcement is to be placed symmetrically.

In Example (a) on page 226 the whole of the concrete section is in compression and the calculation of the top reinforcement is unaltered. An equal area of bottom reinforcement (4-32 mm bars) is provided, which will develop a low compressive stress, unlike the previous unsymmetrical arrangement in which a smaller area of bottom steel was stressed to the maximum.

Example (b) (p. 228) is worked in a similar manner as when calculating unsymmetrical reinforcement. The required areas of top and bottom reinforcement will be found to be equal when $x \simeq 0.87\,d$ i.e. when they are each about 2500 mm^2. Although several trial calculations may be required to obtain a really accurate solution, the area of the top reinforcement is comparatively insensitive to the exact position of the neutral axis and a resonable result may be obtained by assuming $x = d$, by calculating the required area of top reinforcement and by using an equal area of bottom reinforcement.

The example of a portal frame member (p. 229) may be reworked for symmetrical reinforcement as follows, using the modified procedure described above:

Compressive resistance of concrete $\simeq$ external thrust

$$= 300 \text{ kN (67 500 lbf)}.$$

$$x = \frac{300 \times 1000}{0.4 \times 37.5 \times 200} = 100 \text{ mm (4 in)}.$$

Moment of resistance of concrete (about level of bottom reinforcement)

$$= 300\,(412 - 100/2)/1000 = 109 \text{ kN m (965 000 lbf in)}.$$

Moment of external thrust (as before)

$$= 181 \text{ kN m (1 600 000 lbf in)}.$$

Required moment of resistance of top reinforcement

$$= 181 - 109 = 72 \text{ kN m (636 000 lbf in)}.$$

Required compressive resistance of top reinforcement

$$= \frac{72 \times 1000}{375} = 192 \text{ kN (43 200 lbf)}.$$

Strain in top steel

$$= 0.0035 \left(\frac{100 - 32}{100} \right) = 0.0022.$$

From Fig. 4.2 stress in top steel is the maximum design value,

$$= 302 \text{ N/mm}^2 \text{ (43 800 lbf/in}^2).$$

Area of top reinforcement

$$= \frac{192 \times 1000}{302} = 635 \text{ mm}^2 \text{ (0.985 in}^2).$$

Use 2-20 mm bars giving an area of 628 mm^2 (2-$\frac{7}{8}$ in bars give 1.20 in^2). An equal amount of tensile reinforcement is provided at the bottom. Actually the design stress in the tensile reinforcement is

$$0.87 \times 410 = 360 \text{ N/mm}^2 \ (52\ 500 \text{ lbf/in}^2)$$

so that the tensile and compressive forces in the bottom and top reinforcement are not exactly balanced, as initially assumed. In a more exact solution, the depth of the compressive zone (x) would be marginally greater and the required area of the top reinforcement slightly less.

Combined axial load and bending — serviceability limit state

The method of dimensioning described for reinforced concrete members based on the ultimate limit state is also generally applicable to prestressed concrete members. The serviceability limit state would not appear to be critical for reinforced concrete members, and there is in fact little experimental data on the width of cracks in members subject to axial load and bending. However, this limit state must now be considered for Class 1 and 2 prestressed concrete members, in which the limit state is governed by allowable stresses in the concrete, as already discussed in the previous chapter for pure bending, and earlier in the present chapter for axial tension. A method will be described by

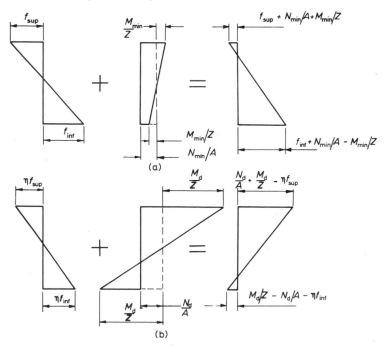

Figure 8.12 Stress conditions for axial force and bending with non-uniform prestress

which the minimum dimensional properties and prestress may be determined for a symmetrical section, and a suitable section selected with the aid of design charts. The solution will depend on whether the prestressing force may act eccentrically or whether it must be axial, thereby producing a uniform prestress in the concrete.

Members with non-uniform prestress

Figure 8.12 represents a symmetrical section of area A and modulus Z in a member which is subjected to an axial force (shown as compressive) and a bending moment. In Fig. 8.12(a) the maximum prestress, that is the prestress at transfer, acts in conjunction with the minimum (sustained) values of the axial force N_{min} and moment M_{min}. In Fig. 8.12(b) the other critical condition is given, in which the prestress has undergone its full reduction to η times its original value and the design working load results in a force N_d and a moment M_d.

If it is specified that the resultant compressive stress in the concrete due to compression and bending at transfer must not exceed $f_{c\,p\,adm}$ and that the flexural tensile stress at the working load must not exceed $f_{t\,adm}$, the following two stress conditions may be written:

$$f_{inf} - \frac{M_{min}}{Z} + \frac{N_{min}}{A} \leqslant f_{c\,p\,adm}, \qquad 8.1$$

$$\frac{M_d}{Z} - \frac{N_d}{A} - \eta f_{inf} \leqslant f_{t\,adm}. \qquad 8.2$$

Eliminating f_{inf} and putting $Z = \alpha b h^2/6$ and $A = \beta bh$, where b and h are the overall breadth and depth of the section

$$\frac{6}{\alpha}\left(\frac{M_d - M_{min}}{bh^2}\right) + \frac{1}{\beta}\left(\frac{N_d - \eta N_{min}}{bh}\right) \leqslant \eta f_{c\,p\,adm} + f_{t\,adm}. \qquad 8.3$$

Similarly, consideration of the stress conditions on the opposite face yields the condition

$$\frac{6}{\alpha}\left(\frac{M_d - \eta M_{min}}{bh^2}\right) + \frac{1}{\beta}\left(\frac{N_d - \eta N_{min}}{bh}\right) \leqslant f_{c\,adm} + \eta f_{t\,p\,adm}, \qquad 8.4$$

where

$f_{c\,adm}$ = allowable compressive stress in concrete under design load,

$f_{t\,p\,adm}$ = allowable flexural tensile stress at transfer of prestress.

The above conditions 8.3 and 8.4 may be represented for a given section as boundary lines on the load-moment diagram. This is the basis of the design charts given in Figs. 8.13 and 8.14, in which the limiting lines are given for a set of symmetrical I sections having a breadth of web equal to three-tenths of the overall breadth of the flange and varying depths of flange. The load and moment scales are in dimensionless units, the allowable stresses being given as fractions of the cube strength of the concrete, and the strength at transfer ($f_{c\ i}$) assumed to be two-thirds of the strength (f_{cu}) when the working load is first applied.

The usefulness of the charts is increased by the introduction of a dimensionless parameter of the form $\psi\ M\ b\ f_{cu}/N^2$ represented on the charts by a family of parabolae which facilitates the preliminary dimensioning of a section, as shown in the following example.

EXAMPLE

A member in a prestressed portal frame is to be designed for a thrust of 90 kN (20 200 lbf) and a bending moment of 88 kN m (870 000 lbf in) at the serviceability state, the minimum thrust and moment being 0.19 of these values.

Assume a cube strength of 40 N/mm² (5800 lbf/in²) at working load; the allowable compressive stress in bending will then be 13.3 N/mm² (1950 lbf/in²) both at transfer and at working load and the allowable tensile stress will be 1 N/mm² (145 lbf/in²) at transfer and 2 N/mm² (290 lbf/in²) at working load. The loss of prestress is estimated to be 20 per cent.

$$M_d - \eta M_{min} = [1 - (0.80 \times 0.19)]\,M_d = 0.85\,M_d.$$

Similarly

$$N_d - \eta M_{min} = 0.85\,N_d.$$

Try

$$b = 200 \text{ mm (8 in)},$$

then

$$\psi = \frac{(M_d - \eta M_{min})bf_{cu}}{(N_d - \eta N_{min})^2} = \frac{0.85 \times 88 \times 10^6 \times 200 \times 40}{(0.85 \times 90 \times 10^3)^2} = 1020$$

From Fig. 8.14, with $\psi = 1020$ and taking $h_f/h = 0.20$

$$\frac{M_d - \eta M_{min}}{bh^2 f_{cu}} \simeq 0.046.$$

Hence

$$h^2 \simeq \frac{0.85 \times 88 \times 10^6}{200 \times 40 \times 0.046} = 203\,000 \text{ mm}^2,$$

$$h \simeq 450 \text{ mm (17.7 in)}.$$

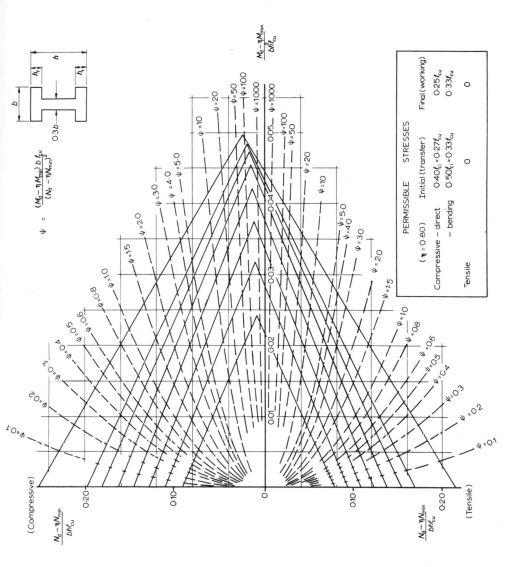

Figure 8.13   Chart for dimensioning section with non-uniform prestress subject to axial load and bending (no tension in concrete)

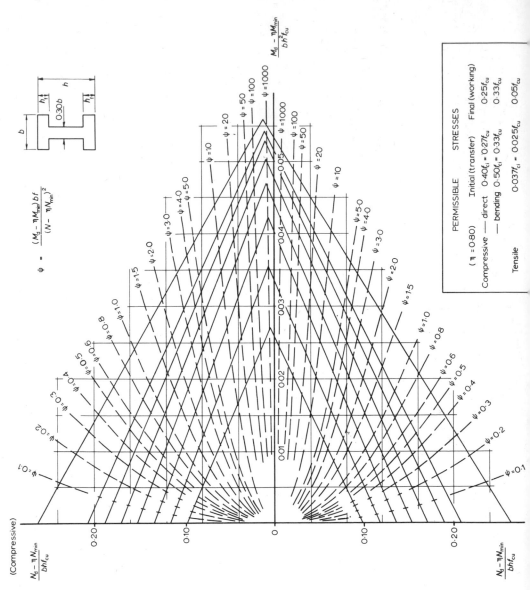

Figure 8.14 Chart for dimensioning sections with non-uniform prestress subject to axial load and bending (tension in concrete)

Suitable dimensions for the section are therefore 450 mm x 200 mm (18 in x 8 in) with a web of breadth 0.3 x 200 = 60 mm (2.4 in) and flanges of average depth 0.20 x 450 = 90 mm (3.6 in). The dimensional properties of the section shown in Fig. 8.15 (a) are

area

$$A = 51\ 500\ \text{mm}^2\ (83.5\ \text{in}^2),$$

second moment of area

$$I = 1275 \times 10^6\ \text{mm}^4\ (3273\ \text{in}^4),$$

section modulus

$$Z = 5.65 \times 10^6\ \text{mm}^3\ (363.7\ \text{in}^3).$$

From equation 8.2 the minimum compressive prestress required at transfer is

$$f_{inf} = \frac{1}{0.80}\left(\frac{88 \times 10^6}{5.65 \times 10^6} - \frac{90 \times 10^3}{51\ 500} - 2\right) = 14.70\ \text{N/mm}^2\ (2000\ \text{lbf/in}^2).$$

Similarly the tensile prestress at transfer must not be greater than

$$1 + \frac{0.19 \times 88 \times 10^6}{5.65 \times 10^6} + \frac{0.19 \times 90 \times 10^3}{51\ 500} = 4.28\ \text{N/mm}^2\ (605\ \text{lbf/in}^2).$$

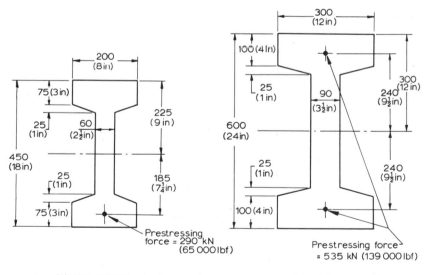

(a) Non-uniform prestress   (b) Uniform prestress

Figure 8.15   Example – Design of prestressed sections for compression and bending

A prestressing force of 290 kN (65 000 lbf) at transfer, in the position indicated in Fig. 8.15(a), will provide a prestress of 15.00 N/mm² (2075 lbf/in²) compression at the bottom and 3.85 N/mm² (520 lbf/in²) tension at the top and is therefore satisfactory.

Members with uniform prestress

A uniform prestress is required in members subject to a bending moment which can act in either of two opposite directions. An addition stress condition illustrated in Fig. 8.16 has now to be considered, since the maximum compressive stress occurs in the extreme case in which the total design load is applied before any reduction of the prestress has taken place.

$$f + \frac{N_d}{A} + \frac{M_d}{Z} \leqslant f_{c\,adm}.$$

8.5

From the equations 8.2 and 8.5

$$\frac{6(1+\eta)M_d}{\alpha\,bh^2} - \frac{(1-\eta)N_d}{\beta\,bh} \leqslant \eta f_{c\,adm} + f_{t\,adm}.$$

8.6

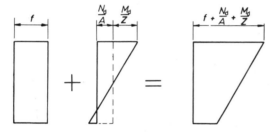

Figure 8.16   Stress conditions for axial force and bending with uniform prestress

When designing for a uniform prestress, the line represented by this expression must be added to the load-moment diagram, as has been done in Fig. 8.17, showing the reduced capacity of a section when a uniform prestress is used in order to accommodate reversal of bending.

EXAMPLE

Rework the previous example using a uniform prestress and allowing no tensile stress in the concrete.

It is found from the position on the chart (Fig. 8.17) of the point representing the values of $N_d/f_{cu}bh$ and $M_d/f_{cu}bh^2$ that the overall dimensions 450 mm x 200 mm (18 in x 8 in) do not permit a solution even with a

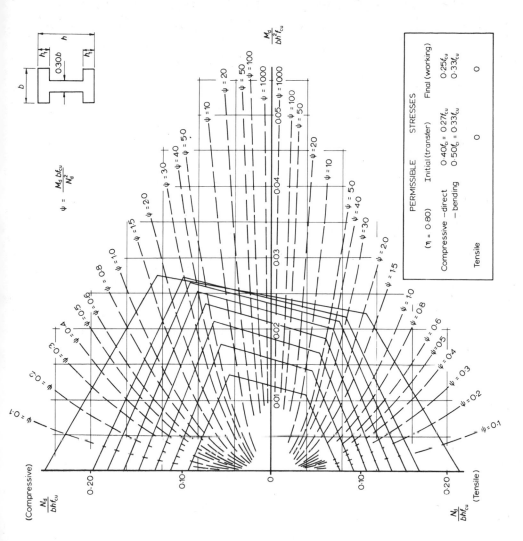

Figure 8.17 Chart for dimensioning sections with uniform prestress subject to axial load and bending (no tension in concrete)

rectangular section. However, using the chart as in the previous example an I-section 600 mm × 300 mm (24 in × 12 in) of the dimensions shown in Fig. 8.15(b) is found to be suitable. The dimensional properties of this section are

$A = 101\ 250\ \text{mm}^2\ (160.5\ \text{in}^2)$,

$I = 4476 \times 10^6\ \text{mm}^4\ (11\ 422\ \text{in}^4)$,

$Z = 14.92 \times 10^6\ \text{mm}^3\ (951.8\ \text{in}^3)$

and the required compressive prestress is found to be 5.28 N/mm² (865 lbf/in²) The prestressing force is therefore $5.28 \times 101\ 250 \times 10^{-3} = 535$ kN (139 000 lbf). In this example the greatly increased cross-sectional area and prestressing force required, demonstrate the poor economy of using a uniform prestress, except where reversal of bending makes this course unavoidable.

The above method of dimensioning prestressed members subject to axial force and bending has been given in greater detail in an earlier paper [8.1].

REFERENCE

[8.1] Bennett, E. W. (Aug. 1966). The design of prestressed members subjected to axial force and bending. *Concr. Construct. Engng,* **61**, 267-274.

9
Shear, torsion and anchorage forces

Dimensions controlled by shear

Shear design formulae may conveniently be expressed in terms of the average shear stress $v = V/b_w d$. The maximum possible ultimate shear resistance of a section is independent of the amount of shear reinforcement, and is attained when there is compressive failure of the concrete, which is mainly governed by the effective depth of the section (d), the web breadth (b_w) and the strength of the concrete.

The condition given for reinforced concrete in the British Code may be written

$$v = \frac{V}{b_w d} \leqslant 0.75 \sqrt{f_{cu}}$$

where

V = ultimate shear, obtained from the characteristic load by applying the specified load factors.

The same condition is applied to prestressed concrete, but in the European Concrete Committee Recommendations the limiting expression is

$$\frac{V - 0.9 P \sin \theta}{b_w d} \leqslant 0.25 \left(\frac{f_c'}{1.5}\right),$$

or approximately

$$\frac{V - 0.9\,P\sin\theta}{b_w\,d} \leqslant 0.13\,f_{cu}.$$

A suitable breadth for the web may be determined from one of the above formulae. Clearly the shear condition will be more critical in a simply supported beam of varying depth, in which d diminishes near the supports, while conversely the effect of haunches in a member of a framed structure will be to increase the shear resistance. It should also be noted that the shear resistance may be improved by a local increase of the web breadth in regions of high shear.

Shear contribution of concrete (British Code)

In the British Code the resistance of a reinforced concrete member without shear reinforcement is represented by an average shear stress v_c which is a function of the percentage of main tensile reinforcement $100\,A_s/b_w d$ and the characteristic strength of the concrete, given by Table 5 in the Code or as Table 5.1 in Chapter 5 (p. 124). It is further assumed that after cracking v_c represents the contribution of the concrete as discussed in Chapter 5. The latter assumption is also made for prestressed concrete, but the cracking load will be different for web shear cracking and for flexure-shear cracking, the lesser of the two values being applicable.

The formula for the average stress v_{co} at which web-shear cracking occurs, is based on the principal tensile stress criterion

$$v_{co} = \frac{V_{co}}{b_w\,d} = 0.67\,\frac{h}{d}\,\sqrt{f_t{}^2 + 0.8\,f_{cp}f_t} + \frac{P\sin\theta}{b_w\,d},$$

where

$$f_t = 0.24\,\sqrt{f_{cu}} = \text{assumed tensile strength of concrete,}$$

f_{cp} = compressive prestress in concrete at level of centroid.

The flexure-shear cracking of a prestressed beam depends on the prior formation of flexural cracks in the region concerned. Neglecting the tensile strength of the concrete, this will occur when the stress becomes zero at the tensile face of the concrete, and allowing a 20% reduction of the prestress the moment M_o is given by

$$M_o = 0.8\,f_{pt}\,I/y$$

where f_{pt} = stress due to prestress only, at tensile face,

y = distance of tensile face from centroid of concrete section,

I = second moment of area of concrete section.

If V is the shear and M the moment for the ultimate limit state at the section considered, the shear corresponding to the moment M_o will be $M_o \, V/M$. It is assumed that flexure-shear cracking occurs after a further increase of shear equal to

$$\left(1 - 0.55 \frac{f_{pe}}{f_{pu}}\right) v_c \, b_w \, d,$$

where

f_{pe} = effective prestress in steel after all losses (not to be greater than $0.6 f_{pu}$),

f_{pu} = tensile strength of steel in tendons,

v_c = shear value of concrete as defined above.

The total average shear stress v_{cr} is therefore given by

$$v_{cr} = \left(1 - 0.55 \frac{f_{pe}}{f_{pu}}\right) v_c + \frac{M_o}{b_w \, d} \frac{V}{M}.$$

This should not be taken as less than $0.1 \sqrt{f_{cu}}$

Calculation of shear reinforcement (British Code)

Under ultimate conditions, the design stress in the shear reinforcement is the characteristic strength (f_{yv}) divided by the factor 1.15. The shear resisted by the reinforcement, according to the conventional formula is therefore:

$$\frac{(0.87 \, f_{yv}) \, d \, A_{sv}}{s \, \sin \alpha}$$

where

A_{sv} = cross-sectional area of the two legs of a stirrup,

s = spacing of stirrups along member,

α = inclination of stirrups to direction of member, to be not less than $45°$.

This is equivalent to an average shear stress on the section of

$$\frac{(0.87 \, f_y) \, d \, A_s}{b_w \, s \, \sin \alpha}.$$

When calculating the required amount of shear reinforcement it is first necessary to check whether the average shear stress v under the design ultimate load exceeds the shear stress contribution of the concrete v_c. If it is less than v_c the resistance of the concrete is sufficient for the whole shear, but a minimum amount of shear reinforcement is specified, unless v is less than one half of v_c. In reinforced concrete the specified minimum ratios are 0.0012 for high yield steel and 0.002 for mild steel. In prestressed concrete the minimum reinforcement should correspond to an average shear stress of 0.4 N/mm^2, i.e.

$$\frac{A_{sv}}{b_w\, s \sin \alpha} \geqslant \frac{0.4\ \text{N/mm}^2}{(0.87\, f_{yv})}$$

If v is greater than v_c the contribution of the concrete to the shear resistance is considered to be equivalent to an average shear stress of v_c so that the amount of shear reinforcement is given by

$$\frac{A_{sv}}{b_w\, s \sin \alpha} = \frac{v - v_c}{(0.87\, f_{yv})}$$

The above expressions are graphically represented in the shear reinforcement charts for high strength steel and mild steel in Figs. 9.1 and 9.2. The maximum

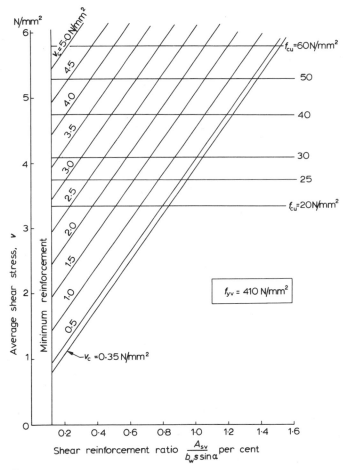

Figure 9.1   Chart for high strength shear reinforcement

shear capacity of the section is represented by the horizontal line for the appropriate value of f_{cu}.

Details of shear reinforcement

Shear reinforcement may consist of stirrups, either perpendicular or inclined to the axis of the member, or of bent-up bars. Perpendicular stirrups are most common and fulfil a useful practical function in enabling the reinforcement to be assembled in a cage before fixing in the formwork or mould. Each crack must be crossed by at least one stirrup if the latter are to function effectively. The spacing should not exceed 0.75 d to ensure this, and a similar limitation of the lateral spacing of the individual legs is imposed by the British Code to ensure an

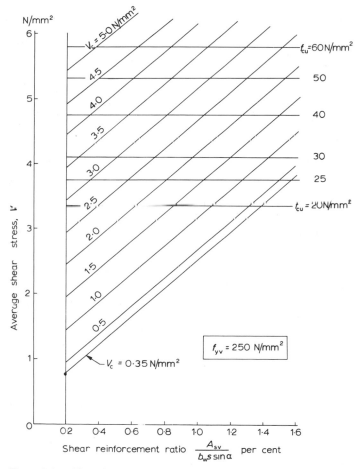

Figure 9.2   Chart for mild steel shear reinforcement

even distribution of stress across the section. It is also necessary that the stirrups should completely enclose the tensile reinforcement in order to prevent failure by splitting along the line of these bars.

Inclined stirrups are not widely used, perhaps on account of their being slightly less convenient to fix. Provided the angle of inclination is not less than 45° their theoretical resistance, in terms of the volumetric percentage of shear reinforcement, is slightly greater than that of perpendicular stirrups depending on the angle of inclination. It has been shown experimentally, however, that the maximum shear resistance of a section is increased when the reinforcement consists of inclined stirrups, and allowance is made for this in the European Concrete Committee's Recommendations. The British Code allows up to 50% of the shear reinforcement to be in the form of bent-up bars, which are convenient in that they may be formed from that part of the main tensile reinforcement which is no longer required in the region of decreasing bending moment and increasing shear. It should, however, be noted that the reduction of the tensile reinforcement may have an adverse effect on the shear resistance by diminishing the value of v_c.

Although the shear resistance developed by bent-up bars is in principle no different to that of inclined stirrups, it has been common practice in this country to design them on the analogy of a lattice girder in which the inclined tensile members are provided by the bent-up bars and inclined compressive members by the resistance of the concrete. The shear resistance is equal to the transverse component of the force developed in the bent-up bars at any section. On this basis the shear resistance of bars bent up at 45° would be equal to the value indicated by the general formula for spacing $s = 2d$, but it is generally recommended that the spacing should not exceed $\sqrt{2}d$.

Only the straight part of each bent-up bar can be considered to be fully effective in resisting shear, and it should be checked that these bars are adequately anchored, using the normal method of calculation. In addition it may be necessary to check the bearing stress at the bends which is given by the expression

$$\frac{\text{Force in bar}}{\text{Internal Radius of bend x size of bar}}$$

According to the British Code this stress should not exceed

$$\frac{1.5\,f_{cu}}{1 + 2d_b/a_b}$$

where d_b is the bar size and a_b should be taken as the clear distance between bars or groups of bars perpendicular to the plane of the bend or, for bars adjacent to the face of the member, the cover plus the bar size.

Where bars are placed in bundles, the bearing stress should be calculated for a single bar of equivalent area.

EXAMPLE – REINFORCED CONCRETE

Calculate the shear reinforcement required in the reinforced concrete T *girder, dimensioned for bending in Chapter 7 (p. 175 ff).*

(a) Stirrup reinforcement alone.

Required ultimate shear at support

$$= 40.4 \times 25 \times 0.5 = 505 \text{ kN (114 000 lbf).}$$

Ultimate average shear stress

$$v = \frac{505 \times 1000}{250 \times 1286} = 1.57 \text{ N/mm}^2 \text{ (228 lbf/in}^2\text{).}$$

Tensile reinforcement ratio

$$\frac{A_s}{b_w d} = \frac{9150}{250 \times 1286} = 2.84\%.$$

$f_{cu} = 45 \text{ N/mm}^2$, hence from Table 5.1 (Code Table 5)

$$v_c = 0.99 \text{ N/mm}^2 \text{ (144 lbf/in}^2\text{).}$$

Shear resistance of concrete alone,

$$V_c = 0.99 \times 250 \times 1286 \times 10^{-3} = 318 \text{ kN (71 500 lbf).}$$

From the shear diagram in Fig. 9.3 it can be seen that the shear resistance of the concrete is greater than the required ultimate shear at distances greater than about 5 m (16 ft) from the support. The minimum shear reinforcement ratio is 0.0012 and the maximum spacing of stirrups is 0.75 x 1286 = 965 mm (38 in). These requirements can be met by 12 mm stirrups at 750 mm centres, so that $A_{sv}/b_w s$ = 226/250 x 750 = 0.00120. At distances greater than about 8.5 m (28 ft) from the support the required ultimate shear is less than $V_c/2$ and the minimum shear reinforcement is no longer required; however, the stirrups have been continued to give support to the longitudinal bars when fixing the reinforcement.

The calculated shear resistance of the above minimum provision of stirrups is

$$\frac{(0.87 \, f_{yv}) A_{sv} d}{s} = \frac{0.87 \times 420 \times 226}{750 \times 1000} = 110 \text{ kN (24 800 lbf).}$$

The combined resistance of the concrete and stirrups is therefore 318 + 110 = 428 kN (96 500 lbf), and from Fig. 9.3(a), additional reinforcement is only necessary over the 2 m (6 ft) nearest to the support. This can be provided

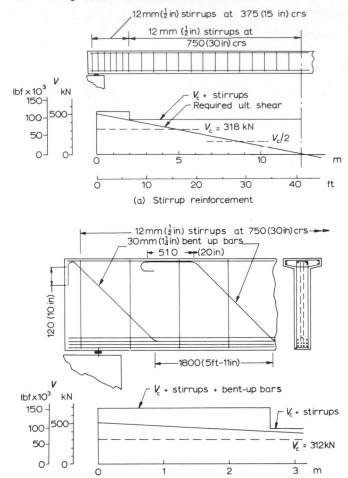

Figure 9.3   Shear reinforcement in reinforced concrete girder

by halving the spacing of the stirrups to 375 mm (15 in), thereby doubling their shear resistance and increasing the total shear resistance to 538 kN (121 000 lbf).

(b) Stirrups and bent-up bars

As an alternative to the increase of the stirrups at the end of the beam, the additional shear resistance can be provided by bending up two of the 30 mm bars at 45°. This will reduce the ratio of tensile reinforcement to 11/13 × 2.84 = 2.40%, decreasing the values of v_c and V_c to 0.97 N/mm² (141 lbf/in²) and 312 kN (70 000 lbf) respectively.

The maximum shear resistance of 30 mm bent-up bars at 1800 mm centres (Fig. 9.3(b)) is

$$\frac{(0.87\,f_{yv})A_{sv}d}{s\sin\alpha} = \frac{0.87 \times 420 \times 705 \times 1286}{1800 \times 0.707 \times 1000} = 260\ \text{kN (58 500 lbf)}.$$

Although this is greater than the shear resistance of the stirrups (110 kN) the resistance actually required from the bent-up bars is only $505 - 428 = 77$ kN which is considerably less.

Alternatively, using the lattice girder method, the maximum spacing of the bent-up bars is $1.41 \times 1286 = 1820$ mm (72 in) and the shear resistance is

$$(0.87\,f_{yv})\,A_{sv}\sin\alpha = 0.87 \times 420 \times 705 \times 0.707 \times 10^{-3}$$
$$= 182\ \text{kN (41 000 lbf)}.$$

Anchorage bond stress for concrete Grade 40 (Code Table 22) with 40% increase for deformed bar

$$= 1.8 \times 1.4 = 2.5\ \text{N/mm}^2\ (363\ \text{lbf/in}^2).$$

Anchorage length (Chapter 2, p. 54)

$$= \frac{0.87 \times 420}{4 \times 2.5} = 37\ \text{diameters}.$$

Allowing 12 diameters for hook in end bend-up bar, anchorage length

$$= (37 - 12) \times 30 = 750\ \text{mm (30 in)}.$$

This may be measured from a point on the bar vertically above the support, so that a length of 250 mm (10 in) is required beyond the end hook. Similarly, allowing 20 diameters for bend and hook in second bent-up bar, the anchorage length is found to be 510 mm (20 in). Allowable bearing stress

$$\frac{1.5\,f_{cu}}{1 + 2\,d_b/a_b} = \frac{1.5 \times 45}{1 + 2 \times 30/110} = 43.7\ \text{N/mm}^2\ (6350\ \text{lbf/in}^2).$$

Actual bearing stress

$$= \frac{0.87 \times 420 \times 705}{r \times 30} = 43.7.$$

internal radius of bend

$$r = 196\ \text{mm, say } 200\ \text{mm (8 in)}.$$

EXAMPLE – PRESTRESSED CONCRETE

Calculate the shear reinforcement required in the prestressed concrete girder dimensioned for bending in Chapter 7 (p. 177 ff).

Shear at support for ultimate limit state

$$V = [1.4\,(7 + 2) + 1.6 \times 15] \times 25 \times 0.5 = 457\ \text{kN (103 000 lbf)}.$$

The shear will be checked at the section 1 m from the support, clear of the end block.

$$V = \frac{11.5}{12.5} \times 457 = 420 \text{ kN (94 700 lbf)}.$$

The inclination of the line of the prestressing force at this point is calculated from the parabolic profile (Fig. 9.4).

Height above lowest point

$$= 370 \times \frac{11.5^2}{12.5^2} = 315 \text{ mm (12.4 in)}.$$

Inclination

$$= \frac{2 \times 315}{11\ 500} = 0.055$$

$$P \sin \theta = 0.85 \times 2430 \times 0.055 = 113 \text{ kN (25 500 lbf)}.$$
$$f_t = 0.24 \sqrt{45} = 1.61 \text{ N/mm}^2 \text{ (234 lbf/in}^2\text{)}.$$

$$f_{cp} = \frac{0.85 \times 2430 \times 1000}{328\ 500} = 6.29 \text{ N/mm}^2 \text{ (910 lbf/in}^2\text{)}.$$

$$V_{co} = \frac{0.67 \times 150 \times 1300}{1000} \sqrt{(1.61^2 + 0.8 \times 1.61 \times 6.29)} + 113$$

$$= 540 \text{ kN (121 000 lbf)}.$$

There will thus be no web-shear cracking, and only the minimum web reinforcement is required.

$$\frac{A_{sv}}{b_w s} \geqslant \frac{0.4}{0.87 \times 420} = 0.109\%.$$

10 mm stirrups at 750 mm crs ($\frac{3}{8}$ in stirrups at 30 in crs) provide 0.139%.

The value of V_{co} only decreases slightly at greater distances from the end of

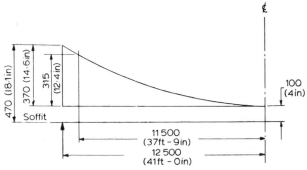

Figure 9.4   Profile of prestressing force

the beam, owing to the slight reduction in the vertical component of the prestressing force, $P \sin \theta$, as shown in Fig. 9.5. It is necessary, however, to check that the shear does not exceed the flexure-shear resistance in the region of the quarter-span point, where there is an appreciable bending moment accompanied by some shear.

At a point 6 m (19.6 ft) from the support the shear

$$V = 237 \text{ kN} (53\ 200 \text{ lbf})$$

and moment

$$M = 2090 \text{ kN m} (18\ 500\ 000 \text{ lbf in}).$$

Eccentricity of prestressing force,

$$e = 655 - \left(\frac{6.5}{12.5}\right)^2 \times 370 = 555 \text{ mm} (21.9 \text{ in}).$$

Effective depth

$$d = 545 + 555 = 1100 \text{ mm} (43.3 \text{ in}).$$

Prestress at bottom

$$f_{\text{pt}} = \frac{0.85 \times 2430 \times 1000}{328\ 500} \left(1 + \frac{555 \times 755}{202\ 000}\right)$$

$$= 19.25 \text{ N/mm}^2 (2800 \text{ lbf/in}^2).$$

$$M_0 = 0.8 \times 19.25 \times \frac{66\ 500 \times 10^6}{755} \times \frac{1}{10^6}$$

$$= 1360 \text{ kN m} (12\ 000\ 000 \text{ lbf in}).$$

$$\frac{100 A_{\text{s}}}{b_{\text{w}} d} = \frac{100 \times 2310}{150 \times 1100} = 1.40.$$

From Table 5.1 (Table 5 in Code),

$$v_{\text{c}} = 0.83 \text{ N/mm}^2 (120 \text{ lbf/in}^2).$$

$$f_{\text{pe}} = 890 \text{ N/mm}^2 (130\ 000 \text{ lbf/in}^2).$$

$$f_{\text{pu}} = 1500 \text{ N/mm}^2 (218\ 000 \text{ lbf/in}^2).$$

$$V_{\text{cr}} = \left(1 - \frac{0.55 \times 890}{1500}\right) 0.83 \times 150 \times \frac{1100}{1000} + 1360 \times \frac{237}{2090}$$

$$= 246 \text{ kN} (55\ 500 \text{ lbf}).$$

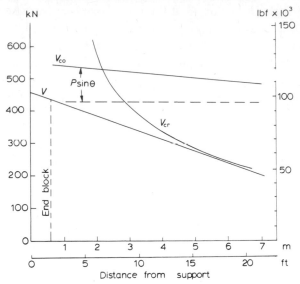

Figure 9.5   Shear diagram for prestressed beam

The curve of flexure to shear strength is shown in Fig. 9.5. In this example the actual shear is less than the strength at all points and the minimum amount of web reinforcement is sufficient.

A possible arrangement of stirrups for a prestressed T or I section is shown in Fig. 9.6. The stirrups may be used to support the draped tendons in the correct positions by means of welded horizontal spacing bars to which the metal sheathing is wired. As an alternative the spacers are sometimes connected to the sides of the mould.

Dimensioning for torsion

Minor torsional effects, such as arise from transverse eccentricity in the loading of members can generally be accommodated by the size of section and amount of reinforcement calculated for bending and shear. If, however, a member is required to resist a major torsional action the dimensions should be determined or checked in accordance with the principles discussed in Chapter 5. This also applies if torsional stiffness has been taken into account in the analysis of the structure.

EXAMPLE

A balcony 2 m (6 ft 6 in) wide, designed for a uniformly distributed load of 4 kN/m² (80 lbf/ft²) is to be cantilevered from a reinforced concrete wall beam,

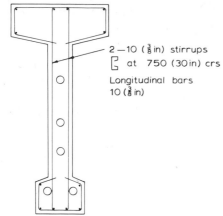

2 —10 ($\frac{3}{8}$ in) stirrups
at 750 (30 in) crs
Longitudinal bars
10 ($\frac{3}{8}$ in)

Figure 9.6   Stirrups in prestressed beam

framing into column faces 4 m (13 ft) apart. Determine the sizes and reinforcement of a suitable slab and beam section (Fig. 9.7).

Estimated overall depth at fixed end of cantilever slab = 1/10 x 2000 = 200 mm (8 in) tapering to 150 mm (6 in) at free end. Characteristic load on balcony (per unit width) is made up by

$$\text{dead load} = \frac{2(0.200 + 0.150)}{2} \times \frac{2400}{1000} \times 9.81 = 8.3 \text{ kN/m (570 lbf/ft)},$$

imposed load = 2 x 4 $\qquad\qquad$ = 8.0 kN/m (550 lbf/ft).

Total load $\qquad\qquad\qquad$ = 16.3 kN/m (1120 lbf/ft).

Ultimate load on balcony
$$= 1.4 \times 8.3 + 1.6 \times 8 = 24.4 \text{ kN/m (1670 lbf/ft)}.$$

Characteristic load on beam (per unit length) is made up by

Balcony
$$= 16.3 \text{ kN/m (1110 lbf/ft)},$$

wall cladding (400 kg/m) and allowance for weight of beam, (400 kg/m)
$$= (400 + 400) \times 9.81/1000 = 7.9 \text{ kN/m (540 lbf/ft)}.$$

ultimate load on beam
$$= 24.4 + 1.4 \times 7.9 = 35.5 \text{ kN/m (2430 lbf/ft)},$$

ultimate shear on beam
$$= 0.5 \times 4.0 \times 35.5 = 71.0 \text{ kN (16 000 lbf)}.$$

Ultimate moment at root of cantilever = torque per unit length of beam

$$= 0.5 \times 2.0 \times 24.4 = 24.4 \text{ kN m } (215\,500 \text{ lbf.in}).$$

Maximum ultimate torque on beam

$$= 0.5 \times 4.0 \times 24.4 = 48.8 \text{ kN m } (431\,000 \text{ lbf.in}).$$

Flexural requirements of wall beam may be met by a section 450 mm by 350 mm (16 in by 14 in) with 2-20 mm (2-¾ in) bottom bars and 2-20 mm (2-¾ in) top bars.

According to British Code flexure shear stress

$$v = \frac{V}{bd} = \frac{71.0 \times 10^3}{350 \times 410} = 0.49 \text{ N/mm}^2 \ (71 \text{ lbf/in}^2).$$

Torsion shear stress

$$v_t = \frac{2T}{h_{min}^2 \left(h_{max} - \dfrac{h_{min}}{3} \right)} = \frac{2 \times 48.8 \times 10^6}{350^2 (450 - 350/3)} = 2.39 \text{ N/mm}^2 \ (347 \text{ lbf/in}^2).$$

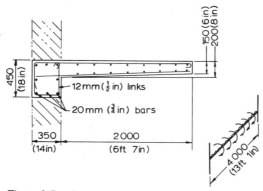

12mm (½ in) links
20mm (¾ in) bars
350 (14in)
2000 (6ft 7in)
150 (6in)
200 (8in)
4000 (13ft 1in)
450 (18in)

Figure 9.7   Example of reinforced concrete beam subject to torsion

If Grade 30 concrete is specified, the allowable ultimate torsion shear stresses from Table 7 of the Code are: $v_{t\,min} = 0.37 \text{ N/mm}^2$ (54 lbf/in²) and $v_{tu} = 4.10 \text{ N/mm}^2$ (596 lbf/in²). Since $v_{t\,min} < v_t < 400/550 \times 4.10$ (400 mm being the larger dimension of a link) and $v + v_t < v_{tu}$, the required torsion resistance may be provided by means of links. Percentage of longitudinal tensile reinforcement

$$= \frac{100\,A_s}{bd} = \frac{100 \times 628}{350 \times 410} = 0.44\%.$$

From Table 5 of the Code, allowable ultimate shear stress in concrete,

$$v_c = 0.40.$$

Links required for shear

$$\frac{A_{sv}}{s_v} \geqslant \frac{b(v - v_c)}{0.87 f_{yv}} = \frac{350\,(0.49 - 0.40)}{0.87 \times 410} = 0.09 \text{ mm}^2/\text{mm } (0.0035 \text{ in}^2/\text{in}).$$

This is less than the minimum requirement of $0.0012\,b$, but the area of links required for combined shear and torsion will be greater than the minimum.

Area of links required for torsion

$$\frac{A_{sv}}{s_v} \geqslant \frac{T}{0.8 x_1 y_1 (0.87 f_{yv})} = \frac{48.8 \times 10^6}{0.8 \times 300 \times 400 \times 0.87 \times 410}$$

$$= 1.43 \text{ mm}^2/\text{mm } (0.0565 \text{ in}^2/\text{in}).$$

Area of total links per unit length

$$\frac{A_{sv}}{s_v} = 0.09 + 1.43 = 1.52 \text{ mm}^2/\text{mm } (0.060 \text{ in}^2/\text{in}).$$

12 mm links at 150 mm centres provide 1.51 mm^2/mm (½ in links at 6 in centres provide 0.065 in^2/in).

Additional longitudinal reinforcement

$$\frac{A_{sl}}{s_v} \left(\frac{f_{yv}}{f_y} \right) (x_1 + y_1) = 1.43 \times 1 \times (300 + 400) = 1001 \text{ mm}^2 \ (1.55 \text{ in}^2).$$

This may be provided by the addition of four 20 mm bars, one at the centre of each leg of the links, increasing the area by 1260 mm^2 (4-¾ in bars = 1.77 in^2) (Fig. 9.7). Alternatively the four existing 20 mm (¾ in) bars at the corners could be replaced by 25 mm (1 in) bars and two 16 mm (⅝ in) bars added at the centres of the larger legs of the links, increasing the area by 1110 mm^2 (1.98 in^2). The distance between the longitudinal bars must not exceed 300 mm (12 in). so that in any solution bars must be placed at the centres of the larger legs.

Anchorage of prestressing tendons

In designing the anchorage arrangements for prestressed tendons the main problem is to accommodate the tensile stresses created by the very large concentrated forces applied to the concrete in the region of the end of the member. The distribution of these forces over the whole cross section, in accordance with the principle of St. Venant, involves a certain length of the member, sometimes termed the 'lead-in' length. Within the end region there exists a complex state of stress in which the most serious effect is a tensile stress (the 'bursting stress') approximately normal to the axis of the member. This tensile stress is frequently sufficient to crack the concrete and, in the absence of adequate reinforcement, can initiate a catastrophic splitting along the whole length of the member.

Assuming that the magnitude and line of action of the total prestressing force has already been determined, the safety and serviceability of the anchorage zone depend on four main considerations.

(a) Design of the anchorage device

Generally this amounts to choosing from a number of well-proved proprietary anchorages, the choice being influenced by cost and suitability for the particular application. The space occupied is often a critical factor, limiting the amount of distribution of the force by the anchorage itself.

In some of the older types of anchorage device the prestressing force is transmitted from the tendon by means of wedges, or a thread and nut, to a steel plate bearing on the end face of the member. It is essential that the distribution plate should be sufficiently stiff, as any appreciable bending will reduce the effective bearing area and increase the concentration of pressure. At present the more common type of anchorage is embedded in the concrete, a more compact arrangement which transmits the force by a combination of bearing at the front of the anchorage and bond at the sides, mainly the latter.

(b) Cross section of the anchorage zone

The severity of the local stresses can often be reduced by increasing the cross-sectional area of the member in the anchorage zone, and it is a common practice to design a solid 'end block' for a member of I or T section, thereby also providing more space for the anchorages. The cross section of the end block may be of rectangular or other appropriate shape as in Fig. 9.8.

The length of the end block should correspond approximately to the 'lead-in' length, and it must be appreciated that, in addition to the redistribution of stresses between the anchorages and the end block section, a further redistribution has to take place between the end block and the flanged section of the main length of the member. The usual length of the end block is between one and one and a quarter times the overall depth.

There has recently been a tendency to dispense with end blocks wherever possible, and to rely on reinforcement to accommodate the local forces. This is

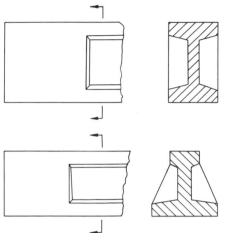

Figure 9.8   Typical end blocks of prestressed beams

not always practicable with post-tensioned tendons because of the space and cover required by each anchorage device, but it can often be done with pre-tensioned tendons since the gradual increase of prestress along the transmission length has the effect of reducing the intensity of stress near the end of the member. The absence of end blocks simplifies construction where one mould is used for members of varying length, and sometimes it is even possible to cut a single long member into shorter lengths.

(c) Arrangement of anchorages

The line of the resultant prestressing force will usually have been more or less predetermined by the bending and shear requirements of the member, but the designer will have a certain degree of freedom in placing the individual anchorages. An important principle, first enunciated by Guyon [9.1] concerns the uniform distribution of the anchorage forces. The local stress conditions in the anchorage zone are least severe when the force applied at each anchorage is exactly opposed by the resistance of a corresponding area of the concrete section at the end of the 'lead-in' length. The line of each force will thus pass through the centroid of the appropriate portion in the diagram of force per unit depth (stress × breadth). Simple examples of this arrangement are shown in Fig. 9.9.

When the anchorage forces are linearly distributed it is only considered necessary to take account of the local stresses arising from the distribution of each anchorage force over its appropriate area. On the other hand non-uniform

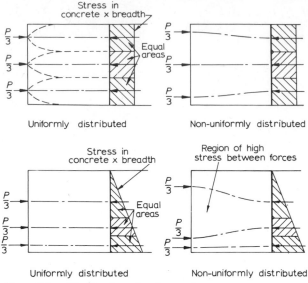

Figure 9.9   Distribution of anchorage forces

distribution may give rise to significant tensile stresses between the anchorage forces as in the example in Fig. 9.9.

(d) Provision of reinforcement

The usual method adopted is to provide sufficient transverse reinforcement in each tensile zone to resist the 'bursting force', obtained by integrating the tensile stresses over the appropriate area of concrete. This is admittedly a rather unrealistic basis of design, because the reinforcement stresses normally assumed can only be developed after cracking of the concrete, which will alter the conditions of homogeneity which have been assumed in calculating the bursting force. The practice has to be accepted, however, in the present absence of a satisfactory method of analysis of a cracked end block.

The tensile stresses and bursting force will be proportional to the anchorage force. For an ultimate limit state calculation according to the British Code the anchorage force should be taken as the jacking load or the ultimate load in the tendon, whichever is greater. If a tendon is effectively bonded by grouting there should be no increase of load at the anchorage even when the beam is loaded to failure; on the contrary the initial force will be reduced with time, mainly by creep of the concrete. On the other hand, when a beam with unbonded tendons is loaded to failure, the anchorage force may be expected to increase to a value approaching the force at other points along the beam. The design stress in the reinforcement is the usual value of $0.87f_y$.

Most of the available data on tensile stresses and bursting forces have been derived for axially loaded prisms. An example of the general distribution of the bursting stress is shown in Figs. 9.10 and 9.11, while Fig. 9.12 gives the total bursting force as a function of the distribution ratio y_{po}/y_o. The results of Zielinski and Rowe [9.2] were obtained experimentally by means of strain measurements on the surface of a concrete block loaded three-dimensionally by means of a prestressing anchorage device, and therefore overestimate the average stresses and total bursting force indicated by the three-dimensional finite element analysis of Yettram and Robbins [9.3]. The bursting forces to be assumed in ultimate limit state design according to the British Code are also shown in Fig. 9.12; these provide a factor (γ_m) of about 1.5 on the strength of the concrete if the values of Yettram and Robbins are accepted as a close approximation of the actual stresses.

The tensile bursting stress commences at a distance between $0.2\,y_o$ and $0.5\,y_o$ from the end of the member, depending on the distribution ratio y_{po}/y_o, and may be considered insignificant at a distance greater than $2\,y_o$. Anchorage reinforcement is therefore usually distributed over the length extending from $0.2\,y_o$ to $2\,y_o$ from the loaded face as recommended in the British Code. This length and the total bursting force may be slightly reduced by deducting the region in which the bursting stress is less than the tensile strength of the concrete, but the refinement is usually unwarranted having regard to the small amount of reinforcement involved.

The reinforcement must be calculated in the two principal transverse

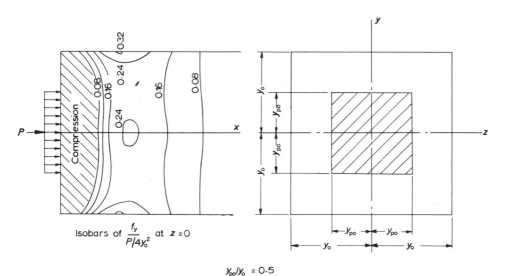

Isobars of $\dfrac{f_y}{P/4y_o^2}$ at $z=0$

$y_{po}/y_o = 0.5$

Figure 9.10   Isobar diagram of bursting stress (Yettram and Robbins [9.3]

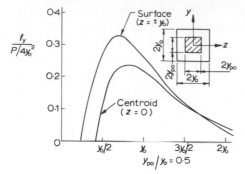

Figure 9.11   Distribution of bursting stress, $y_{po}/y_o = 0.5$ (Yettram and Robbins [9.3]

directions (i.e. vertical and horizontal for a horizontal beam), but one system of parallel bars may often be arranged to resist the bursting forces in several anchorage zones. Anchorage reinforcement has sometimes been designed as a set of meshes, but the commonest arrangement is a system of closed stirrups. If desired the reinforcement may be distributed in proportion to the intensity of the bursting stress, but in view of the degree of approximation in the analysis, a highly sophisticated reinforcement system is not usually justified.

EXAMPLE

Use the method recommended in the British Code to dimension a suitable system of anchorage reinforcement for the prestressed beam, for which the shear calculations have been given earlier in the chapter.

The five tendons are first arranged so that the anchorage forces are linearly distributed (Fig. 9.13). It can be assumed that under these conditions the lead-in length will be less than half the length of the anchorage block, and the prestress

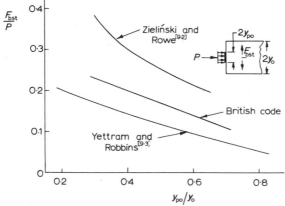

Figure 9.12   Total bursting force on axially loaded square prisms

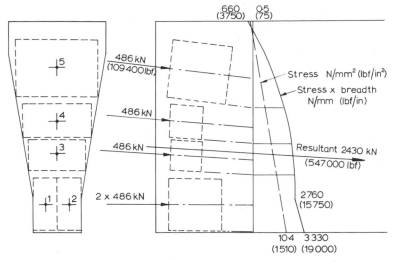

Figure 9.13   Arrangement of anchorage forces and equivalent prisms

has therefore been calculated for the cross section 600 mm (24 in) from the end. The diagram of force per unit depth is drawn by multiplying the prestress diagram by the breadth of the section and the area of this diagram is divided into portions which are proportional to the magnitude of the anchorage forces. The tendons are then arranged so that the anchorage forces pass through the centroids of their respective portions in the diagram of force per unit depth. If this is correctly done the position of the resultant prestressing force should be unchanged. At this stage it is desirable to draw the anchorage devices, with a suitable recess which will be filled with cement mortar to seal the ends of the tendons after prestressing (Fig. 9.14).

The imaginary prisms, on which the individual anchorage forces are considered to act, may now be constructed. This has been done in Fig. 9.13 which shows how the boundaries of the prisms are defined by the faces of the anchorage block or by the boundaries between the portions in the diagram of force per unit depth. The bursting forces on each prism are then calculated. For example, considering tendons 1 and 2 together

prestressing force

$$P = 972 \text{ kN } (218\,800 \text{ lbf}),$$

vertical distribution ration

$$y_{po}/y_o = 60/160 = 0.38,$$

vertical bursting force (Table 45 of British Code)

$$F_{bst} = 0.21 \times 972 = 204 \text{ kN } (46\,000 \text{ lbf}).$$

This force acts in a region from 0.2 x 160 = 32 mm (1.3 in) to 2 x 160 = 320 mm (12.5 in) from the end of the prism, i.e. 112 mm (4.4 in) to 400 mm (15.8 in) from the end of the beam.

Average force per unit length

$$= \frac{204\,000}{320 - 32} = 710 \text{ N/mm (4060 lbf/in)}.$$

Horizontal distribution ratio

$$z_{po}/z_o = 125/150 = 0.83.$$

Referring to Table 45 of the Code, this ratio is seen to be sufficiently large for the bursting force to be neglected.

The bursting forces are similarly calculated for all the anchorages, and the results are given in Table 9.1.

If the anchorage reinforcement is to consist of 10 mm ($\frac{3}{8}$ in) steel with a characteristic tensile strength of 420 N/mm^2 (61 000 lbf/in^2) the design ultimate resistance of one stirrup will be

$$0.87 \times 420 \times 2 \times 78.5 = 57\,400 \text{ N (12 900 lbf)}.$$

A possible arrangement of anchorage reinforcement is shown in Fig. 9.15. The closed rectangular stirrups at 75 mm (3 in) centres develop a resistance of 57 400/75 = 765 N/mm (4380 lbf/in) for anchorages 1 and 2, while the remaining anchorages are each reinforced by pairs of horizontal and vertical bars at 150 mm (6 in) centres. This reinforcement is not required to resist anchorage

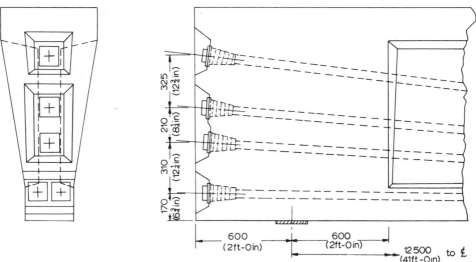

Figure 9.14   Details of anchorage block

Table 9.1 *Bursting forces at anchorages of prestressed beam*

Tendon no.	Vertical distribution					Horizontal distribution			
	Prestressing force P_k kN (lbf)	Dist'n ratio y_{po}/y_o	Bursting force $F_{bst\,y}$ kN (lbf)	Limits of bursting force mm (in) from end of beam	Force per unit length N/mm (lbf/in)	Dist'n ratio z_{po}/z_o	Bursting force $F_{bst\,z}$ kN (lbf)	Limits of bursting force mm (in) from end of beam	Force per unit length N/mm (lbf/in)
1 & 2	972 (218 800)	0.21	204 (46 000)	112 400 (4.4) (15.8)	710 (4060)	0.83	—	—	—
3	486 (109 400)	0.67	59 (13 300)	120 280 (4.7) (11.0)	360 (2060)	0.34	116 (26 100)	135 450 (5.3) (17.7)	370 (2120)
4	486 (109 400)	0.60	68 (15 300)	110 290 (4.3) (11.4)	380 (2170)	0.29	112 (25 200)	130 510 (5.1) (20.1)	300 (1710)
5	486 (109 400)	0.33	107 (24 100)	130 450 (5.1) (17.7)	330 (1890)	0.24	126 (28 300)	140 590 (5.5) (23.2)	280 (1600)

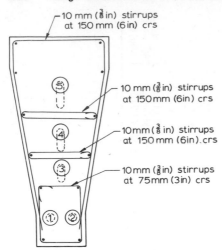

Figure 9.15   Anchorage reinforcement

bursting forces at a greater distance than 590 mm (23.2 in) from the end of the beam, but it would be advisable to continue all the stirrups at a spacing of about 150 mm (6 in) to the end of the anchorage block in order to accommodate the local stresses due to the change of cross section.

REFERENCES

[9.1] Guyon, Y. (1953) *Prestressed Concrete*, Ch. 6. London: Contractors Record.

[9.2] Zielinski, J. and Rowe, R. E. (1960) An investigation of the stress distribution in the anchorage zones of post-tensioned concrete members. *Research Report* No. 9, London: Cement and Concrete Association.

[9.3] Yettram, A. L. and Robbins, K. (June 1969) Anchorage zone stresses in axially post-tensioned members of uniform rectangular section. *Mag. Concr. Res.*, 21, 103-112.

Appendix A

Computation of dimensional properties of a prestressed concrete section

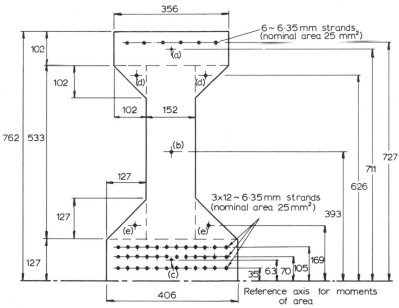

Figure A.1   Section divided for computation of dimensional properties

Part of section	Dimensions mm	A mm^2	y^* mm	Ay mm$^3 \times 10^3$	Ay^2 mm$^4 \times 10^6$	I mm$^4 \times 10^6$
(a)	356 102	36 312	711	25 817.382	18 356.478	31.482
(b)	152 533	81 016	393	31 874.658	12 526.741	1 917.980
(c)	406 127	51 512	63	3 248.406	204.650	69.304
(d)	2/½/102 102	10 404	626	6 512.904	4 077.078	6.013
(e)	2/½/127 127	16 129	169	2 725.801	460.660	14.452
Σ		195 423		70 179.601	35 625.607	2 039.231
						35 625.607
						37 664.838

$$\bar{y} = \frac{\Sigma Ay}{\Sigma A} \qquad 359$$

$(\Sigma A)\bar{y}^2 = (\Sigma Ay)\bar{y}$ 25 194.477

I (centroid)   12 470.361

	A	y	Ay	Ay²	
	12/25	300	70	21.000	1.470
	12/25	300	105	31.500	3.307
	6/25	150	727	109.050	79.279
Σ		1.050		172.050	84.423
Nominal section	195.423	359	70 179.601		→ 37 664.838
Tendons	1.050		172.050	84.423	→ 84.423
Concrete section (deducting area of tendons†)	194 373	360	70 007.551		37 580.415
$\bar{y} = \dfrac{\Sigma Ay}{\Sigma A}$					
$(\Sigma A)\bar{y}^2 = (\Sigma Ay)\bar{y}$					25 202.718
I (centroid)					12 377.697
Blank section	195 423	359	70 179.601		→ 37 664.838
equivalent additional concrete area of tendons ($\alpha = 6$)	1050 × 5 = 5 250		860.250		422.115 → 422.115
(blank section equivalent additional concrete area of tendons)	200 673		71 039.851		38 002.530
Equivalent (transformed) section $\bar{y} = \dfrac{\Sigma Ay}{\Sigma A}$		354			
(including tendons†) $(\Sigma A)\bar{y}^2 = (\Sigma Ay)\bar{y}$					25 148.107
I (centroid)					12 854.423

---

* For slide-rule computation, it is more accurate to take an arbitrary reference axis near the centroid.

† The equivalent additional concrete area of any non-prestressed reinforcement must be included in both the concrete section and the equivalent section.

Appendix B
Table of section modulus ratios

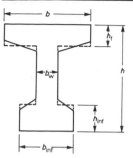

Section modulus ratios

$$\frac{Z_{inf}}{bh^2/6}$$

$$\frac{Z_{sup}}{bh^2/6}$$

$\dfrac{b_w}{b}$	$\dfrac{b_{inf}}{b}$	$\dfrac{h_{inf}}{h}$	$\dfrac{h_f}{h}=0.10$	$\dfrac{h_f}{h}=0.15$	$\dfrac{h_f}{h}=0.20$	$\dfrac{h_f}{h}=0.25$	$\dfrac{h_f}{h}=0.30$	$\dfrac{h_f}{h}=0.35$
0.10	0.10							
		0.10	0.151	0.153	0.152	0.152	0.153	0.157
			0.377	0.445	0.475	0.480	0.473	0.463
		0.15	0.151	0.153	0.152	0.152	0.153	0.157
			0.377	0.445	0.475	0.480	0.473	0.463
		0.20	0.151	0.153	0.152	0.152	0.153	0.157
			0.377	0.445	0.475	0.480	0.473	0.463
		0.25	0.151	0.153	0.152	0.152	0.153	0.157
			0.377	0.445	0.475	0.480	0.473	0.463
		0.30	0.151	0.153	0.152	0.152	0.153	0.157
			0.377	0.445	0.475	0.480	0.473	0.463
		0.35	0.151	0.153	0.152	0.152	0.153	0.157
			0.377	0.445	0.475	0.480	0.473	0.463
	0.15							
		0.10	0.174	0.175	0.175	0.174	0.175	0.179
			0.398	0.473	0.510	0.519	0.514	0.504
		0.15	0.182	0.184	0.184	0.183	0.184	0.187
			0.404	0.482	0.521	0.532	0.528	0.518
		0.20	0.190	0.192	0.191	0.191	0.191	0.194
			0.408	0.489	0.529	0.542	0.539	0.529
		0.25	0.196	0.198	0.198	0.197	0.197	0.200
			0.411	0.493	0.535	0.549	0.547	0.538
		0.30	0.201	0.204	0.203	0.202	0.202	0.205
			0.412	0.495	0.539	0.554	0.552	0.544
		0.35	0.205	0.208	0.207	0.206	0.207	0.209
			0.412	0.496	0.541	0.557	0.556	0.548
	0.20							
		0.10	0.196	0.198	0.197	0.196	0.197	0.200
			0.416	0.498	0.540	0.554	0.551	0.541
		0.15	0.213	0.216	0.215	0.214	0.214	0.217
			0.426	0.513	0.559	0.576	0.575	0.566
		0.20	0.228	0.231	0.230	0.229	0.229	0.231
			0.432	0.522	0.572	0.592	0.593	0.585
		0.25	0.240	0.243	0.243	0.241	0.241	0.243
			0.435	0.528	0.581	0.603	0.606	0.599
		0.30	0.249	0.253	0.253	0.251	0.251	0.253
			0.436	0.531	0.586	0.610	0.614	0.608
		0.35	0.257	0.261	0.261	0.260	0.259	0.260
			0.436	0.531	0.588	0.614	0.619	0.614

Section modulus ratios

$$\frac{Z_{inf}}{bh^2/6}$$

$$\frac{Z_{sup}}{bh^2/6}$$

$\dfrac{b_w}{b}$	$\dfrac{b_{inf}}{b}$	$\dfrac{h_{inf}}{h}$	$\dfrac{h_f}{h}=0.10$	$\dfrac{h_f}{h}=0.15$	$\dfrac{h_f}{h}=0.20$	$\dfrac{h_f}{h}=0.25$	$\dfrac{h_f}{h}=0.30$	$\dfrac{h_f}{h}=0.35$
0.10	0.25							
		0.10	0.218	0.220	0.219	0.218	0.218	0.221
			0.432	0.520	0.567	0.585	0.584	0.575
		0.15	0.244	0.247	0.246	0.244	0.244	0.246
			0.444	0.538	0.592	0.614	0.617	0.609
		0.20	0.265	0.269	0.268	0.267	0.266	0.268
			0.451	0.549	0.607	0.634	0.640	0.634
		0.25	0.283	0.287	0.287	0.285	0.284	0.285
			0.454	0.555	0.617	0.647	0.655	0.651
		0.30	0.296	0.302	0.302	0.300	0.299	0.299
			0.454	0.557	0.621	0.654	0.664	0.661
		0.35	0.306	0.313	0.313	0.312	0.310	0.311
			0.453	0.557	0.623	0.657	0.669	0.668
	0.30							
		0.10	0.240	0.242	0.241	0.240	0.240	0.242
			0.445	0.539	0.591	0.613	0.615	0.607
		0.15	0.274	0.278	0.277	0.275	0.274	0.276
			0.459	0.560	0.619	0.647	0.654	0.648
		0.20	0.302	0.307	0.306	0.304	0.303	0.304
			0.466	0.571	0.636	0.669	0.679	0.676
		0.25	0.325	0.331	0.331	0.328	0.327	0.327
			0.468	0.577	0.646	0.682	0.696	0.695
		0.30	0.342	0.349	0.350	0.348	0.346	0.346
			0.468	0.578	0.649	0.689	0.705	0.706
		0.35	0.354	0.363	0.364	0.363	0.361	0.360
			0.466	0.577	0.650	0.691	0.709	0.712
	0.35							
		0.10	0.262	0.265	0.264	0.262	0.261	0.263
			0.457	0.556	0.613	0.638	0.642	0.636
		0.15	0.304	0.308	0.308	0.305	0.304	0.305
			0.472	0.578	0.643	0.676	0.686	0.683
		0.20	0.339	0.344	0.344	0.342	0.340	0.340
			0.478	0.589	0.661	0.699	0.714	0.714
		0.25	0.366	0.373	0.374	0.371	0.369	0.369
			0.480	0.594	0.669	0.712	0.731	0.733
		0.30	0.386	0.396	0.397	0.395	0.392	0.391
			0.479	0.595	0.672	0.718	0.739	0.744
		0.35	0.400	0.412	0.414	0.412	0.410	0.409
			0.476	0.592	0.672	0.719	0.743	0.749

Section modulus ratios

$$\frac{Z_{inf}}{bh^2/6}$$

$$\frac{Z_{sup}}{bh^2/6}$$

$\dfrac{b_w}{b}$	$\dfrac{b_{inf}}{b}$	$\dfrac{h_{inf}}{h}$	$\dfrac{h_f}{h}=0.10$	$\dfrac{h_f}{h}=0.15$	$\dfrac{h_f}{h}=0.20$	$\dfrac{h_f}{h}=0.25$	$\dfrac{h_f}{h}=0.30$	$\dfrac{h_f}{h}=0.35$
0.10	0.40							
		0.10	0.283	0.287	0.286	0.284	0.283	0.284
			0.468	0.571	0.632	0.661	0.668	0.663
		0.15	0.334	0.339	0.338	0.336	0.334	0.335
			0.483	0.594	0.664	0.702	0.715	0.714
		0.20	0.375	0.382	0.381	0.379	0.377	0.376
			0.489	0.605	0.682	0.725	0.744	0.747
		0.25	0.406	0.415	0.416	0.414	0.411	0.410
			0.490	0.609	0.689	0.738	0.761	0.766
		0.30	0.429	0.441	0.443	0.441	0.438	0.436
			0.488	0.608	0.691	0.743	0.769	0.777
		0.35	0.445	0.460	0.463	0.461	0.459	0.456
			0.485	0.605	0.690	0.743	0.771	0.781
	0.45							
		0.10	0.305	0.309	0.308	0.306	0.304	0.306
			0.477	0.584	0.650	0.682	0.692	0.688
		0.15	0.364	0.369	0.369	0.366	0.364	0.364
			0.492	0.607	0.683	0.725	0.742	0.743
		0.20	0.410	0.418	0.418	0.416	0.413	0.412
			0.497	0.618	0.699	0.748	0.771	0.776
		0.25	0.445	0.457	0.458	0.456	0.452	0.451
			0.498	0.621	0.706	0.759	0.787	0.796
		0.30	0.471	0.486	0.488	0.486	0.483	0.481
			0.495	0.620	0.707	0.763	0.794	0.806
		0.35	0.488	0.506	0.511	0.509	0.506	0.504
			0.492	0.616	0.705	0.763	0.795	0.809
	0.50							
		0.10	0.327	0.331	0.330	0.327	0.326	0.327
			0.486	0.596	0.666	0.702	0.714	0.712
		0.15	0.393	0.399	0.399	0.396	0.393	0.393
			0.500	0.620	0.699	0.745	0.765	0.769
		0.20	0.445	0.455	0.455	0.452	0.449	0.447
			0.505	0.629	0.715	0.768	0.795	0.803
		0.25	0.484	0.497	0.500	0.497	0.494	0.491
			0.505	0.631	0.721	0.778	0.810	0.822
		0.30	0.512	0.529	0.533	0.531	0.528	0.525
			0.502	0.629	0.721	0.781	0.816	0.831
		0.35	0.529	0.551	0.557	0.556	0.553	0.550
			0.498	0.625	0.718	0.780	0.816	0.833

	Section modulus ratios	$\dfrac{Z_{\text{inf}}}{bh^2/6}$
		$\dfrac{Z_{\text{sup}}}{bh^2/6}$

$\dfrac{b_w}{b}$	$\dfrac{b_{\text{inf}}}{b}$	$\dfrac{h_{\text{inf}}}{h}$	$\dfrac{h_f}{h}=0.10$	$\dfrac{h_f}{h}=0.15$	$\dfrac{h_f}{h}=0.20$	$\dfrac{h_f}{h}=0.25$	$\dfrac{h_f}{h}=0.30$	$\dfrac{h_f}{h}=0.35$
0.10	0.55							
		0.10	0.348	0.353	0.352	0.349	0.347	0.348
			0.494	0.607	0.680	0.720	0.734	0.734
		0.15	0.422	0.429	0.429	0.426	0.423	0.422
			0.507	0.630	0.714	0.763	0.787	0.793
		0.20	0.480	0.491	0.492	0.489	0.485	0.483
			0.512	0.639	0.729	0.786	0.816	0.827
		0.25	0.522	0.538	0.540	0.538	0.534	0.531
			0.511	0.641	0.734	0.795	0.830	0.845
		0.30	0.551	0.572	0.577	0.576	0.572	0.568
			0.507	0.638	0.733	0.797	0.835	0.853
		0.35	0.570	0.595	0.603	0.602	0.599	0.595
			0.504	0.633	0.729	0.795	0.834	0.855
	0.60							
		0.10	0.370	0.375	0.374	0.371	0.369	0.369
			0.500	0.618	0.694	0.736	0.753	0.754
		0.15	0.451	0.459	0.459	0.456	0.452	0.451
			0.514	0.640	0.727	0.780	0.807	0.815
		0.20	0.514	0.526	0.528	0.525	0.521	0.518
			0.517	0.648	0.741	0.802	0.835	0.849
		0.25	0.559	0.577	0.581	0.579	0.574	0.571
			0.516	0.649	0.745	0.810	0.848	0.866
		0.30	0.590	0.613	0.620	0.619	0.615	0.611
			0.512	0.645	0.744	0.811	0.852	0.873
		0.35	0.609	0.637	0.647	0.648	0.644	0.640
			0.508	0.640	0.739	0.808	0.851	0.874
	0.65							
		0.10	0.391	0.397	0.395	0.393	0.390	0.390
			0.507	0.627	0.706	0.752	0.771	0.774
		0.15	0.480	0.489	0.489	0.485	0.482	0.479
			0.520	0.649	0.739	0.795	0.825	0.835
		0.20	0.547	0.561	0.563	0.560	0.556	0.553
			0.523	0.656	0.752	0.816	0.852	0.869
		0.25	0.596	0.616	0.621	0.619	0.614	0.610
			0.521	0.656	0.755	0.823	0.865	0.885
		0.30	0.628	0.654	0.662	0.662	0.658	0.653
			0.517	0.652	0.753	0.823	0.867	0.891
		0.35	0.646	0.679	0.691	0.692	0.689	0.684
			0.513	0.647	0.748	0.819	0.865	0.891

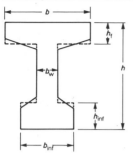

Section modulus ratios

$$\frac{Z_{inf}}{bh^2/6}$$

$$\frac{Z_{sup}}{bh^2/6}$$

$\frac{b_w}{b}$	$\frac{b_{inf}}{b}$	$\frac{h_{inf}}{h}$	$\frac{h_f}{h}=0.10$	$\frac{h_f}{h}=0.15$	$\frac{h_f}{h}=0.20$	$\frac{h_f}{h}=0.25$	$\frac{h_f}{h}=0.30$	$\frac{h_f}{h}=0.35$
0.10	0.70							
		0.10	0.413	0.418	0.417	0.414	0.411	0.410
			0.513	0.635	0.718	0.766	0.788	0.792
		0.15	0.508	0.518	0.518	0.515	0.511	0.508
			0.525	0.657	0.750	0.809	0.842	0.854
		0.20	0.581	0.596	0.599	0.596	0.591	0.587
			0.527	0.663	0.762	0.829	0.868	0.887
		0.25	0.632	0.654	0.660	0.658	0.654	0.649
			0.525	0.662	0.764	0.835	0.879	0.903
		0.30	0.665	0.694	0.704	0.704	0.700	0.695
			0.521	0.658	0.761	0.834	0.881	0.907
		0.35	0.683	0.720	0.733	0.735	0.733	0.728
			0.517	0.652	0.756	0.830	0.878	0.906
	0.75							
		0.10	0.434	0.440	0.439	0.436	0.433	0.431
			0.518	0.643	0.728	0.779	0.803	0.809
		0.15	0.537	0.548	0.548	0.544	0.540	0.537
			0.530	0.664	0.760	0.822	0.857	0.872
		0.20	0.614	0.631	0.634	0.631	0.626	0.622
			0.531	0.669	0.771	0.841	0.883	0.904
		0.25	0.667	0.692	0.699	0.698	0.693	0.688
			0.528	0.668	0.772	0.846	0.893	0.918
		0.30	0.701	0.734	0.745	0.746	0.742	+0.737
			0.524	0.663	0.769	0.844	0.894	+0.922
		0.35	0.719	0.759	0.775	0.778	0.776	0.771
			0.520	0.658	0.763	0.839	0.890	0.920
	0.80							
		0.10	0.455	0.462	0.461	0.457	0.454	0.452
			0.523	0.650	0.738	0.792	0.818	0.826
		0.15	0.565	0.577	0.577	0.574	0.569	0.565
			0.534	0.670	0.769	0.834	0.871	0.888
		0.20	0.646	0.665	0.669	0.666	0.661	0.656
			0.535	0.675	0.779	0.851	0.896	0.919
		0.25	0.702	0.729	0.737	0.736	0.732	0.726
			0.532	0.673	0.780	0.856	0.905	0.933
		0.30	0.736	0.772	0.785	0.787	0.783	0.778
			0.527	0.668	0.776	0.853	0.905	0.936
		0.35	0.753	0.798	0.816	0.820	0.818	0.813
			0.524	0.662	0.770	0.848	0.901	0.933

Section modulus ratios

$$\frac{Z_{inf}}{bh^2/6}$$

$$\frac{Z_{sup}}{bh^2/6}$$

$\dfrac{b_w}{b}$	$\dfrac{b_{inf}}{b}$	$\dfrac{h_{inf}}{h}$	$\dfrac{h_f}{h}=0.10$	$\dfrac{h_f}{h}=0.15$	$\dfrac{h_f}{h}=0.20$	$\dfrac{h_f}{h}=0.25$	$\dfrac{h_f}{h}=0.30$	$\dfrac{h_f}{h}=0.35$
0.10	0.85							
		0.10	0.476	0.483	0.482	0.479	0.475	0.473
			0.527	0.657	0.747	0.803	0.832	0.841
		0.15	0.592	0.606	0.607	0.603	0.598	0.594
			0.538	0.676	0.777	0.845	0.885	0.903
		0.20	0.678	0.699	0.704	0.701	0.695	0.690
			0.538	0.680	0.787	0.861	0.908	0.934
		0.25	0.736	0.766	0.775	0.775	0.770	0.764
			0.535	0.678	0.787	0.865	0.916	0.946
		0.30	0.770	0.810	0.825	0.827	0.824	0.818
			0.531	0.672	0.782	0.862	0.916	0.948
		0.35	0.787	0.836	0.856	0.861	0.860	0.855
			0.527	0.667	0.776	0.856	0.911	0.945
	0.90							
		0.10	0.497	0.505	0.504	0.500	0.496	0.494
			0.532	0.663	0.756	0.814	0.845	0.856
		0.15	0.620	0.634	0.636	0.632	0.626	0.622
			0.541	0.682	0.785	0.855	0.897	0.918
		0.20	0.710	0.732	0.738	0.735	0.730	0.724
			0.542	0.685	0.794	0.870	0.920	0.947
		0.25	0.769	0.802	0.813	0.813	0.808	0.802
			0.538	0.682	0.793	0.873	0.927	0.959
		0.30	0.804	0.847	0.864	0.867	0.864	0.858
			0.533	0.676	0.788	0.869	0.925	0.960
		0.35	0.820	0.872	0.895	0.902	0.900	0.896
			0.530	0.671	0.781	0.863	0.920	0.956
	0.95							
		0.10	0.518	0.526	0.526	0.522	0.517	0.515
			0.536	0.669	0.764	0.824	0.857	0.870
		0.15	0.648	0.663	0.665	0.661	0.655	0.650
			0.545	0.687	0.792	0.864	0.908	0.931
		0.20	0.741	0.766	0.772	0.770	0.764	0.758
			0.544	0.689	0.800	0.879	0.930	0.960
		0.25	0.802	0.838	0.850	0.850	0.846	0.839
			0.540	0.686	0.798	0.880	0.936	0.970
		0.30	0.837	0.884	0.902	0.906	0.903	0.897
			0.536	0.680	0.793	0.876	0.934	0.971
		0.35	0.852	0.908	0.933	0.942	0.941	0.936
			0.533	0.675	0.786	0.870	0.929	0.966

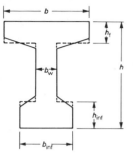

Section modulus ratios

$$\dfrac{Z_{inf}}{bh^2/6}$$

$$\dfrac{Z_{sup}}{bh^2/6}$$

$\dfrac{b_w}{b}$	$\dfrac{b_{inf}}{b}$	$\dfrac{h_{inf}}{h}$	$\dfrac{h_f}{h}=0.10$	$\dfrac{h_f}{h}=0.15$	$\dfrac{h_f}{h}=0.20$	$\dfrac{h_f}{h}=0.25$	$\dfrac{h_f}{h}=0.30$	$\dfrac{h_f}{h}=0.35$
0.10	1.00							
		0.10	0.539	0.548	0.547	0.543	0.538	0.535
			0.539	0.675	0.772	0.834	0.869	0.883
		0.15	0.675	0.691	0.693	0.689	0.684	0.678
			0.548	0.691	0.799	0.873	0.919	0.944
		0.20	0.772	0.799	0.806	0.804	0.798	0.791
			0.547	0.693	0.806	0.886	0.940	0.971
		0.25	0.834	0.873	0.886	0.887	0.883	0.876
			0.543	0.689	0.804	0.887	0.945	0.981
		0.30	0.869	0.919	0.940	0.945	0.942	0.936
			0.538	0.684	0.798	0.883	0.942	0.981
		0.35	0.883	0.944	0.971	0.981	0.981	0.976
			0.535	0.678	0.791	0.876	0.936	0.976
0.15	0.15							
		0.10	0.213	0.217	0.217	0.216	0.216	0.218
			0.419	0.496	0.539	0.556	0.557	0.551
		0.15	0.213	0.217	0.217	0.216	0.216	0.218
			0.419	0.496	0.539	0.556	0.557	0.551
		0.20	0.213	0.217	0.217	0.216	0.216	0.218
			0.419	0.496	0.539	0.556	0.557	0.551
		0.25	0.213	0.217	0.217	0.216	0.216	0.218
			0.419	0.496	0.539	0.556	0.557	0.551
		0.30	0.213	0.217	0.217	0.216	0.216	0.218
			0.419	0.496	0.539	0.556	0.557	0.551
		0.35	0.213	0.217	0.217	0.216	0.216	0.218
			0.419	0.496	0.539	0.556	0.557	0.551
	0.20							
		0.10	0.234	0.239	0.239	0.238	0.238	0.239
			0.435	0.517	0.564	0.585	0.588	0.582
		0.15	0.243	0.248	0.248	0.247	0.246	0.248
			0.440	0.523	0.572	0.594	0.598	0.593
		0.20	0.250	0.255	0.255	0.254	0.253	0.255
			0.444	0.528	0.578	0.601	0.606	0.602
		0.25	0.255	0.261	0.261	0.260	0.259	0.260
			0.445	0.531	0.582	0.606	0.612	0.608
		0.30	0.260	0.266	0.266	0.265	0.264	0.265
			0.446	0.533	0.585	0.610	0.616	0.612
		0.35	0.263	0.269	0.270	0.269	0.268	0.269
			0.446	0.533	0.586	0.611	0.618	0.615

Section modulus ratios
$$\frac{Z_{inf}}{bh^2/6}$$
$$\frac{Z_{sup}}{bh^2/6}$$

$\frac{b_w}{b}$	$\frac{b_{inf}}{b}$	$\frac{h_{inf}}{h}$	$\frac{h_f}{h}=0.10$	$\frac{h_f}{h}=0.15$	$\frac{h_f}{h}=0.20$	$\frac{h_f}{h}=0.25$	$\frac{h_f}{h}=0.30$	$\frac{h_f}{h}=0.35$
0.15	0.25							
		0.10	0.256	0.261	0.261	0.259	0.259	0.260
			0.450	0.536	0.587	0.611	0.616	0.611
		0.15	0.272	0.278	0.278	0.276	0.276	0.277
			0.458	0.547	0.601	0.628	0.635	0.631
		0.20	0.286	0.292	0.292	0.291	0.290	0.291
			0.464	0.555	0.611	0.640	0.649	0.646
		0.25	0.297	0.303	0.304	0.303	0.301	0.302
			0.466	0.559	0.617	0.648	0.658	0.656
		0.30	0.305	0.313	0.314	0.312	0.311	0.311
			0.467	0.561	0.621	0.652	0.664	0.663
		0.35	0.312	0.320	0.321	0.320	0.318	0.319
			0.467	0.561	0.622	0.655	0.667	0.667
	0.30							
		0.10	0.277	0.282	0.282	0.281	0.280	0.281
			0.463	0.553	0.608	0.635	0.642	0.639
		0.15	0.301	0.307	0.308	0.306	0.305	0.306
			0.474	0.568	0.627	0.657	0.668	0.666
		0.20	0.321	0.328	0.329	0.328	0.326	0.326
			0.480	0.577	0.639	0.673	0.685	0.685
		0.25	0.337	0.345	0.347	0.345	0.343	0.343
			0.483	0.582	0.646	0.682	0.697	0.698
		0.30	0.350	0.359	0.360	0.359	0.357	0.357
			0.484	0.584	0.650	0.687	0.704	0.706
		0.35	0.359	0.369	0.371	0.370	0.368	0.367
			0.483	0.583	0.650	0.690	0.707	0.711
	0.35							
		0.10	0.298	0.303	0.304	0.302	0.301	0.302
			0.475	0.568	0.627	0.657	0.666	0.664
		0.15	0.330	0.337	0.338	0.336	0.334	0.335
			0.487	0.586	0.649	0.684	0.697	0.697
		0.20	0.357	0.365	0.366	0.364	0.362	0.362
			0.494	0.596	0.663	0.701	0.718	0.720
		0.25	0.377	0.387	0.389	0.387	0.385	0.384
			0.497	0.601	0.671	0.712	0.731	0.735
		0.30	0.393	0.404	0.406	0.405	0.403	0.402
			0.497	0.602	0.674	0.717	0.738	0.743
		0.35	0.404	0.417	0.420	0.419	0.417	0.415
			0.496	0.601	0.674	0.718	0.741	0.748

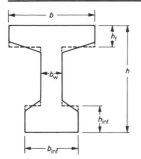

Section modulus ratios

$$\frac{Z_{\text{inf}}}{bh^2/6}$$

$$\frac{Z_{\text{sup}}}{bh^2/6}$$

$\dfrac{b_{\text{w}}}{b}$	$\dfrac{b_{\text{inf}}}{b}$	$\dfrac{h_{\text{inf}}}{h}$	$\dfrac{h_{\text{f}}}{h}=0.10$	$\dfrac{h_{\text{f}}}{h}=0.15$	$\dfrac{h_{\text{f}}}{h}=0.20$	$\dfrac{h_{\text{f}}}{h}=0.25$	$\dfrac{h_{\text{f}}}{h}=0.30$	$\dfrac{h_{\text{f}}}{h}=0.35$
0.15	0.40							
		0.10	0.319	0.325	0.325	0.324	0.322	0.323
			0.486	0.582	0.644	0.677	0.689	0.688
		0.15	0.359	0.367	0.367	0.366	0.364	0.363
			0.499	0.602	0.669	0.708	0.724	0.726
		0.20	0.391	0.401	0.402	0.400	0.398	0.397
			0.506	0.612	0.684	0.726	0.747	0.751
		0.25	0.416	0.428	0.430	0.428	0.426	0.425
			0.509	0.617	0.691	0.737	0.760	0.767
		0.30	0.435	0.449	0.452	0.451	0.448	0.446
			0.508	0.618	0.694	0.742	0.767	0.776
		0.35	0.448	0.464	0.468	0.467	0.465	0.463
			0.506	0.616	0.693	0.743	0.769	0.780
	0.45							
		0.10	0.339	0.346	0.347	0.345	0.343	0.343
			0.495	0.595	0.660	0.696	0.710	0.710
		0.15	0.387	0.396	0.397	0.395	0.393	0.392
			0.510	0.616	0.687	0.729	0.748	0.753
		0.20	0.426	0.436	0.438	0.436	0.434	0.432
			0.517	0.626	0.702	0.749	0.772	0.779
		0.25	0.455	0.468	0.471	0.470	0.467	0.465
			0.519	0.631	0.709	0.759	0.786	0.796
		0.30	0.476	0.492	0.496	0.495	0.492	0.490
			0.518	0.631	0.711	0.763	0.792	0.804
		0.35	0.491	0.509	0.515	0.514	0.512	0.509
			0.515	0.628	0.710	0.763	0.794	0.807
	0.50							
		0.10	0.360	0.367	0.368	0.366	0.364	0.364
			0.504	0.607	0.675	0.713	0.730	0.731
		0.15	0.416	0.425	0.426	0.424	0.422	0.421
			0.519	0.629	0.704	0.749	0.771	0.777
		0.20	0.460	0.472	0.474	0.472	0.469	0.467
			0.526	0.639	0.719	0.769	0.795	0.805
		0.25	0.493	0.508	0.512	0.510	0.507	0.505
			0.527	0.643	0.725	0.779	0.809	0.821
		0.30	0.517	0.535	0.540	0.539	0.536	0.534
			0.526	0.642	0.726	0.782	0.814	0.829
		0.35	0.532	0.554	0.561	0.561	0.558	0.555
			0.523	0.639	0.724	0.781	0.815	0.832

Section modulus ratios: $\dfrac{Z_{inf}}{bh^2/6}$, $\dfrac{Z_{sup}}{bh^2/6}$

$\dfrac{b_w}{b}$	$\dfrac{b_{inf}}{b}$	$\dfrac{h_{inf}}{h}$	$\dfrac{h_f}{h}=0.10$	$\dfrac{h_f}{h}=0.15$	$\dfrac{h_f}{h}=0.20$	$\dfrac{h_f}{h}=0.25$	$\dfrac{h_f}{h}=0.30$	$\dfrac{h_f}{h}=0.35$
0.15	0.55							
		0.10	0.381	0.389	0.389	0.387	0.385	0.385
			0.513	0.618	0.689	0.730	0.748	0.751
		0.15	0.444	0.454	0.456	0.454	0.451	0.449
			0.528	0.640	0.718	0.766	0.791	0.799
		0.20	0.493	0.507	0.509	0.508	0.504	0.502
			0.534	0.650	0.733	0.787	0.816	0.828
		0.25	0.530	0.547	0.552	0.550	0.547	0.544
			0.535	0.653	0.739	0.796	0.829	0.844
		0.30	0.556	0.577	0.583	0.583	0.580	0.576
			0.533	0.652	0.739	0.798	0.834	0.852
		0.35	0.572	0.597	0.605	0.606	0.603	0.600
			0.530	0.649	0.737	0.797	0.834	0.853
	0.60							
		0.10	0.402	0.410	0.411	0.408	0.406	0.405
			0.520	0.628	0.702	0.745	0.765	0.770
		0.15	0.472	0.483	0.485	0.483	0.480	0.478
			0.535	0.651	0.732	0.783	0.810	0.820
		0.20	0.526	0.541	0.545	0.543	0.539	0.537
			0.541	0.660	0.746	0.803	0.835	0.850
		0.25	0.567	0.586	0.591	0.590	0.587	0.583
			0.541	0.662	0.751	0.811	0.847	0.865
		0.30	0.594	0.618	0.625	0.626	0.622	0.619
			0.539	0.661	0.751	0.813	0.852	0.872
		0.35	0.612	0.639	0.649	0.651	0.648	0.644
			0.536	0.657	0.748	0.811	0.851	0.873
	0.65							
		0.10	0.422	0.431	0.432	0.430	0.427	0.426
			0.527	0.638	0.714	0.759	0.782	0.788
		0.15	0.500	0.512	0.514	0.512	0.508	0.506
			0.542	0.660	0.744	0.798	0.828	0.840
		0.20	0.559	0.576	0.580	0.578	0.574	0.571
			0.548	0.669	0.758	0.817	0.852	0.869
		0.25	0.603	0.624	0.630	0.629	0.626	0.622
			0.547	0.671	0.762	0.825	0.864	0.884
		0.30	0.632	0.658	0.667	0.668	0.665	0.661
			0.545	0.668	0.761	0.826	0.867	0.890
		0.35	0.650	0.680	0.692	0.695	0.692	0.688
			0.541	0.664	0.757	0.823	0.866	0.890

Section modulus ratios

$$\frac{Z_{inf}}{bh^2/6}$$

$$\frac{Z_{sup}}{bh^2/6}$$

$\dfrac{b_w}{b}$	$\dfrac{b_{inf}}{b}$	$\dfrac{h_{inf}}{h}$	$\dfrac{h_f}{h}=0.10$	$\dfrac{h_f}{h}=0.15$	$\dfrac{h_f}{h}=0.20$	$\dfrac{h_f}{h}=0.25$	$\dfrac{h_f}{h}=0.30$	$\dfrac{h_f}{h}=0.35$
0.15	0.70							
		0.10	0.443	0.452	0.453	0.451	0.448	0.446
			0.534	0.647	0.725	0.773	0.797	0.804
		0.15	0.527	0.540	0.543	0.540	0.537	0.534
			0.549	0.669	0.755	0.812	0.844	0.858
		0.20	0.592	0.610	0.614	0.613	0.609	0.605
			0.553	0.677	0.769	0.831	0.868	0.887
		0.25	0.638	0.661	0.669	0.668	0.665	0.660
			0.553	0.678	0.772	0.838	0.879	0.901
		0.30	0.669	0.698	0.708	0.709	0.706	0.702
			0.550	0.675	0.771	0.838	0.882	0.906
		0.35	0.686	0.721	0.734	0.738	0.735	0.731
			0.546	0.671	0.766	0.835	0.880	0.906
	0.75							
		0.10	0.463	0.473	0.474	0.472	0.469	0.467
			0.540	0.655	0.735	0.785	0.811	0.820
		0.15	0.555	0.569	0.571	0.569	0.565	0.562
			0.555	0.677	0.765	0.824	0.859	0.875
		0.20	0.624	0.643	0.649	0.647	0.643	0.639
			0.559	0.685	0.778	0.843	0.883	0.904
		0.25	0.673	0.698	0.707	0.707	0.703	0.699
			0.558	0.685	0.781	0.849	0.893	0.917
		0.30	0.705	0.736	0.748	0.750	0.747	0.743
			0.554	0.682	0.779	0.849	0.895	0.922
		0.35	0.722	0.760	0.776	0.780	0.778	0.774
			0.551	0.677	0.774	0.845	0.892	0.920
	0.80							
		0.10	0.484	0.494	0.495	0.493	0.490	0.487
			0.545	0.663	0.745	0.797	0.825	0.836
		0.15	0.582	0.597	0.600	0.598	0.594	0.590
			0.560	0.684	0.775	0.836	0.873	0.890
		0.20	0.655	0.677	0.683	0.681	0.677	0.673
			0.564	0.691	0.787	0.854	0.896	0.919
		0.25	0.707	0.735	0.745	0.745	0.741	0.736
			0.562	0.691	0.789	0.860	0.906	0.932
		0.30	0.740	0.775	0.788	0.791	0.788	0.783
			0.559	0.687	0.787	0.858	0.907	0.935
		0.35	0.757	0.799	0.816	0.821	0.820	0.816
			0.555	0.683	0.782	0.854	0.903	0.933

Section modulus ratios

$$\frac{Z_{inf}}{bh^2/6}$$

$$\frac{Z_{sup}}{bh^2/6}$$

$\dfrac{b_w}{b}$	$\dfrac{b_{inf}}{b}$	$\dfrac{h_{inf}}{h}$	$\dfrac{h_f}{h}=0.10$	$\dfrac{h_f}{h}=0.15$	$\dfrac{h_f}{h}=0.20$	$\dfrac{h_f}{h}=0.25$	$\dfrac{h_f}{h}=0.30$	$\dfrac{h_f}{h}=0.35$
0.15	0.85							
		0.10	0.504	0.515	0.516	0.514	0.510	0.508
			0.551	0.670	0.754	0.808	0.838	0.850
		0.15	0.609	0.625	0.629	0.626	0.622	0.618
			0.565	0.691	0.784	0.847	0.886	0.905
		0.20	0.687	0.710	0.716	0.715	0.711	0.706
			0.568	0.697	0.795	0.864	0.909	0.933
		0.25	0.741	0.771	0.782	0.783	0.779	0.774
			0.566	0.697	0.797	0.869	0.917	0.946
		0.30	0.774	0.812	0.827	0.831	0.828	0.823
			0.562	0.693	0.794	0.867	0.918	0.948
		0.35	0.792	0.836	0.856	0.862	0.861	0.857
			0.559	0.688	0.788	0.863	0.914	0.946
	0.90							
		0.10	0.524	0.536	0.537	0.535	0.531	0.528
			0.556	0.676	0.763	0.819	0.850	0.864
		0.15	0.636	0.653	0.657	0.655	0.650	0.646
			0.569	0.697	0.792	0.858	0.898	0.919
		0.20	0.718	0.742	0.750	0.749	0.745	0.740
			0.572	0.703	0.803	0.874	0.920	0.947
		0.25	0.774	0.807	0.819	0.820	0.816	0.811
			0.570	0.702	0.804	0.878	0.928	0.958
		0.30	0.808	0.849	0.866	0.870	0.868	0.863
			0.566	0.698	0.800	0.876	0.928	0.960
		0.35	0.825	0.873	0.895	0.902	0.902	0.897
			0.562	0.693	0.795	0.870	0.923	0.957
	0.95							
		0.10	0.545	0.557	0.558	0.556	0.552	0.549
			0.560	0.683	0.771	0.829	0.862	0.877
		0.15	0.662	0.681	0.685	0.683	0.678	0.674
			0.574	0.703	0.800	0.867	0.910	0.932
		0.20	0.748	0.775	0.783	0.783	0.778	0.773
			0.576	0.708	0.810	0.883	0.931	0.959
		0.25	0.806	0.842	0.855	0.857	0.854	0.848
			0.573	0.707	0.810	0.886	0.938	0.970
		0.30	0.841	0.885	0.904	0.909	0.907	0.902
			0.569	0.702	0.806	0.883	0.937	0.971
		0.35	0.857	0.909	0.933	0.942	0.942	0.938
			0.566	0.697	0.800	0.878	0.932	0.967

Section modulus ratios

$$\frac{Z_{inf}}{bh^2/6} \qquad \frac{Z_{sup}}{bh^2/6}$$

$\dfrac{b_w}{b}$	$\dfrac{b_{inf}}{b}$	$\dfrac{h_{inf}}{h}$	$\dfrac{h_f}{h}=0.10$	$\dfrac{h_f}{h}=0.15$	$\dfrac{h_f}{h}=0.20$	$\dfrac{h_f}{h}=0.25$	$\dfrac{h_f}{h}=0.30$	$\dfrac{h_f}{h}=0.35$
0.15	1.00							
		0.10	0.565	0.577	0.579	0.576	0.572	0.569
			0.565	0.689	0.779	0.839	0.873	0.889
		0.15	0.689	0.708	0.713	0.711	0.706	0.701
			0.577	0.708	0.807	0.876	0.921	0.945
		0.20	0.779	0.807	0.816	0.816	0.812	0.806
			0.579	0.713	0.816	0.891	0.941	0.971
		0.25	0.839	0.876	0.891	0.894	0.890	0.885
			0.576	0.711	0.816	0.894	0.947	0.981
		0.30	0.873	0.921	0.941	0.947	0.946	0.940
			0.572	0.706	0.812	0.890	0.946	0.981
		0.35	0.889	0.945	0.971	0.981	0.981	0.977
			0.569	0.701	0.806	0.885	0.940	0.977
0.20	0.20							
		0.10	0.270	0.277	0.278	0.277	0.276	0.277
			0.457	0.536	0.585	0.610	0.618	0.616
		0.15	0.270	0.277	0.278	0.277	0.276	0.277
			0.457	0.536	0.585	0.610	0.618	0.616
		0.20	0.270	0.277	0.278	0.277	0.276	0.277
			0.457	0.536	0.585	0.610	0.618	0.616
		0.25	0.270	0.277	0.278	0.277	0.276	0.277
			0.457	0.536	0.585	0.610	0.618	0.616
		0.30	0.270	0.277	0.278	0.277	0.276	0.277
			0.457	0.536	0.585	0.610	0.618	0.616
		0.35	0.270	0.277	0.278	0.277	0.276	0.277
			0.457	0.536	0.585	0.610	0.618	0.616
	0.25							
		0.10	0.291	0.298	0.299	0.298	0.297	0.298
			0.471	0.553	0.605	0.633	0.642	0.641
		0.15	0.299	0.306	0.308	0.307	0.305	0.306
			0.475	0.559	0.612	0.640	0.651	0.650
		0.20	0.305	0.313	0.314	0.313	0.312	0.312
			0.478	0.562	0.616	0.646	0.657	0.657
		0.25	0.310	0.319	0.320	0.319	0.318	0.318
			0.480	0.565	0.619	0.649	0.661	0.661
		0.30	0.315	0.323	0.325	0.324	0.323	0.323
			0.480	0.566	0.621	0.652	0.664	0.665
		0.35	0.318	0.326	0.328	0.327	0.326	0.326
			0.480	0.566	0.622	0.653	0.666	0.667

Section modulus ratios: $\dfrac{Z_{inf}}{bh^2/6}$ $\dfrac{Z_{sup}}{bh^2/6}$

$\dfrac{b_w}{b}$	$\dfrac{b_{inf}}{b}$	$\dfrac{h_{inf}}{h}$	$\dfrac{h_f}{h}=0.10$	$\dfrac{h_f}{h}=0.15$	$\dfrac{h_f}{h}=0.20$	$\dfrac{h_f}{h}=0.25$	$\dfrac{h_f}{h}=0.30$	$\dfrac{h_f}{h}=0.35$
0.20	0.30							
		0.10	0.311	0.319	0.320	0.319	0.318	0.318
			0.484	0.569	0.624	0.654	0.665	0.665
		0.15	0.327	0.335	0.337	0.336	0.334	0.334
			0.491	0.578	0.635	0.667	0.681	0.681
		0.20	0.340	0.349	0.350	0.349	0.348	0.348
			0.496	0.585	0.643	0.677	0.691	0.693
		0.25	0.350	0.360	0.362	0.361	0.359	0.359
			0.498	0.588	0.648	0.683	0.699	0.701
		0.30	0.358	0.368	0.371	0.370	0.368	0.368
			0.499	0.590	0.651	0.687	0.703	0.707
		0.35	0.364	0.375	0.378	0.377	0.375	0.374
			0.499	0.590	0.652	0.688	0.706	0.710
	0.35							
		0.10	0.331	0.340	0.341	0.340	0.339	0.338
			0.495	0.583	0.641	0.673	0.687	0.688
		0.15	0.355	0.364	0.366	0.365	0.363	0.363
			0.505	0.596	0.657	0.692	0.708	0.710
		0.20	0.374	0.384	0.386	0.385	0.383	0.383
			0.511	0.604	0.667	0.704	0.722	0.726
		0.25	0.389	0.400	0.403	0.402	0.400	0.399
			0.514	0.608	0.673	0.712	0.731	0.737
		0.30	0.400	0.412	0.416	0.415	0.413	0.412
			0.514	0.610	0.676	0.716	0.737	0.743
		0.35	0.409	0.422	0.426	0.425	0.423	0.422
			0.514	0.610	0.676	0.718	0.739	0.746
	0.40							
		0.10	0.352	0.360	0.362	0.361	0.359	0.359
			0.506	0.597	0.657	0.692	0.707	0.709
		0.15	0.383	0.393	0.395	0.394	0.392	0.391
			0.518	0.612	0.676	0.714	0.733	0.737
		0.20	0.408	0.419	0.422	0.421	0.419	0.417
			0.524	0.621	0.688	0.729	0.750	0.756
		0.25	0.427	0.440	0.443	0.442	0.440	0.439
			0.527	0.626	0.694	0.737	0.760	0.768
		0.30	0.442	0.456	0.460	0.460	0.457	0.456
			0.527	0.627	0.697	0.742	0.766	0.775
		0.35	0.452	0.468	0.473	0.473	0.470	0.469
			0.526	0.626	0.697	0.743	0.768	0.778

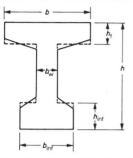

Section modulus ratios $\dfrac{Z_{\text{inf}}}{bh^2/6}$

$\dfrac{Z_{\text{sup}}}{bh^2/6}$

$\dfrac{b_w}{b}$	$\dfrac{b_{\text{inf}}}{b}$	$\dfrac{h_{\text{inf}}}{h}$	$\dfrac{h_f}{h}=0.10$	$\dfrac{h_f}{h}=0.15$	$\dfrac{h_f}{h}=0.20$	$\dfrac{h_f}{h}=0.25$	$\dfrac{h_f}{h}=0.30$	$\dfrac{h_f}{h}=0.35$
0.20	0.45							
		0.10	0.372	0.381	0.383	0.382	0.380	0.379
			0.516	0.609	0.672	0.709	0.726	0.729
		0.15	0.410	0.421	0.424	0.422	0.420	0.419
			0.529	0.627	0.694	0.735	0.755	0.762
		0.20	0.441	0.454	0.457	0.456	0.454	0.452
			0.536	0.636	0.707	0.751	0.774	0.783
		0.25	0.465	0.479	0.483	0.483	0.480	0.478
			0.539	0.641	0.713	0.760	0.786	0.796
		0.30	0.482	0.499	0.504	0.504	0.501	0.499
			0.539	0.642	0.715	0.764	0.791	0.803
		0.35	0.495	0.513	0.519	0.519	0.517	0.515
			0.537	0.640	0.715	0.764	0.793	0.806
	0.50							
		0.10	0.392	0.402	0.404	0.403	0.401	0.400
			0.525	0.621	0.686	0.725	0.744	0.749
		0.15	0.438	0.449	0.452	0.451	0.449	0.447
			0.539	0.640	0.710	0.754	0.777	0.784
		0.20	0.474	0.488	0.492	0.491	0.488	0.486
			0.546	0.650	0.723	0.771	0.797	0.807
		0.25	0.502	0.518	0.523	0.523	0.520	0.518
			0.549	0.654	0.730	0.780	0.808	0.821
		0.30	0.522	0.541	0.547	0.547	0.545	0.542
			0.548	0.654	0.731	0.783	0.814	0.828
		0.35	0.536	0.557	0.564	0.565	0.563	0.560
			0.546	0.653	0.730	0.783	0.815	0.830
	0.55							
		0.10	0.412	0.422	0.425	0.423	0.421	0.420
			0.534	0.632	0.699	0.740	0.761	0.767
		0.15	0.465	0.478	0.481	0.480	0.477	0.475
			0.549	0.652	0.724	0.771	0.796	0.806
		0.20	0.507	0.522	0.526	0.526	0.523	0.521
			0.556	0.662	0.738	0.789	0.817	0.830
		0.25	0.539	0.557	0.562	0.562	0.559	0.557
			0.558	0.666	0.744	0.797	0.829	0.844
		0.30	0.561	0.582	0.589	0.590	0.587	0.584
			0.557	0.665	0.746	0.800	0.834	0.850
		0.35	0.576	0.599	0.608	0.610	0.608	0.605
			0.555	0.663	0.744	0.800	0.834	0.852

Section modulus ratios

$$\frac{Z_{inf}}{bh^2/6}$$

$$\frac{Z_{sup}}{bh^2/6}$$

$\dfrac{b_w}{b}$	$\dfrac{b_{inf}}{b}$	$\dfrac{h_{inf}}{h}$	$\dfrac{h_f}{h}=0.10$	$\dfrac{h_f}{h}=0.15$	$\dfrac{h_f}{h}=0.20$	$\dfrac{h_f}{h}=0.25$	$\dfrac{h_f}{h}=0.30$	$\dfrac{h_f}{h}=0.35$
0.20	0.60							
		0.10	0.432	0.443	0.445	0.444	0.442	0.440
			0.542	0.642	0.711	0.754	0.777	0.784
		0.15	0.492	0.506	0.509	0.508	0.505	0.503
			0.557	0.663	0.738	0.787	0.814	0.826
		0.20	0.539	0.556	0.561	0.560	0.557	0.555
			0.564	0.673	0.752	0.805	0.836	0.851
		0.25	0.575	0.594	0.601	0.601	0.598	0.595
			0.566	0.676	0.758	0.813	0.847	0.864
		0.30	0.599	0.622	0.631	0.632	0.629	0.626
			0.564	0.675	0.758	0.816	0.852	0.871
		0.35	0.615	0.641	0.652	0.654	0.652	0.649
			0.562	0.673	0.756	0.814	0.851	0.872
	0.65							
		0.10	0.452	0.463	0.466	0.465	0.462	0.460
			0.549	0.651	0.723	0.768	0.792	0.800
		0.15	0.519	0.534	0.538	0.536	0.534	0.531
			0.565	0.673	0.750	0.802	0.831	0.844
		0.20	0.572	0.589	0.595	0.595	0.591	0.588
			0.572	0.683	0.764	0.820	0.853	0.870
		0.25	0.610	0.632	0.639	0.640	0.637	0.633
			0.573	0.685	0.769	0.828	0.864	0.883
		0.30	0.637	0.662	0.672	0.673	0.671	0.667
			0.571	0.684	0.769	0.829	0.868	0.889
		0.35	0.653	0.682	0.694	0.697	0.695	0.692
			0.568	0.681	0.767	0.827	0.867	0.889
	0.70							
		0.10	0.472	0.484	0.487	0.485	0.483	0.481
			0.556	0.660	0.734	0.781	0.806	0.816
		0.15	0.546	0.562	0.566	0.565	0.562	0.559
			0.572	0.682	0.762	0.815	0.847	0.861
		0.20	0.603	0.623	0.629	0.629	0.625	0.622
			0.579	0.692	0.775	0.833	0.869	0.887
		0.25	0.645	0.669	0.677	0.678	0.675	0.671
			0.579	0.694	0.780	0.841	0.879	0.901
		0.30	0.673	0.701	0.712	0.714	0.712	0.708
			0.577	0.692	0.780	0.842	0.882	0.906
		0.35	0.690	0.722	0.736	0.740	0.738	0.735
			0.574	0.689	0.776	0.839	0.881	0.905

Section modulus ratios	$\dfrac{Z_{inf}}{bh^2/6}$
	$\dfrac{Z_{sup}}{bh^2/6}$

$\dfrac{b_w}{b}$	$\dfrac{b_{inf}}{b}$	$\dfrac{h_{inf}}{h}$	$\dfrac{h_f}{h}=0.10$	$\dfrac{h_f}{h}=0.15$	$\dfrac{h_f}{h}=0.20$	$\dfrac{h_f}{h}=0.25$	$\dfrac{h_f}{h}=0.30$	$\dfrac{h_f}{h}=0.35$
0.20	0.75							
		0.10	0.492	0.504	0.507	0.506	0.503	0.501
			0.563	0.669	0.744	0.793	0.820	0.831
		0.15	0.573	0.589	0.594	0.593	0.590	0.586
			0.579	0.691	0.773	0.828	0.861	0.878
		0.20	0.635	0.656	0.663	0.662	0.659	0.655
			0.585	0.700	0.786	0.846	0.884	0.904
		0.25	0.680	0.705	0.715	0.716	0.713	0.709
			0.585	0.702	0.790	0.853	0.894	0.917
		0.30	0.709	0.740	0.752	0.755	0.753	0.749
			0.583	0.700	0.789	0.853	0.896	0.921
		0.35	0.727	0.762	0.777	0.782	0.780	0.777
			0.580	0.696	0.785	0.850	0.894	0.920
	0.80							
		0.10	0.512	0.525	0.528	0.526	0.523	0.521
			0.569	0.676	0.754	0.804	0.833	0.845
		0.15	0.599	0.617	0.622	0.621	0.618	0.614
			0.585	0.699	0.783	0.840	0.875	0.893
		0.20	0.666	0.688	0.696	0.696	0.693	0.689
			0.591	0.707	0.795	0.857	0.897	0.919
		0.25	0.713	0.741	0.752	0.753	0.751	0.746
			0.591	0.709	0.799	0.864	0.907	0.931
		0.30	0.744	0.777	0.791	0.795	0.793	0.789
			0.588	0.706	0.797	0.864	0.908	0.935
		0.35	9.762	0.800	0.817	0.823	0.822	0.818
			0.585	0.702	0.793	0.860	0.906	0.934
	0.85							
		0.10	0.531	0.545	0.548	0.547	0.544	0.541
			0.575	0.684	0.763	0.815	0.845	0.858
		0.15	0.626	0.644	0.650	0.649	0.645	0.642
			0.591	0.706	0.792	0.851	0.888	0.907
		0.20	0.697	0.721	0.729	0.730	0.726	0.722
			0.596	0.714	0.804	0.868	0.910	0.933
		0.25	0.747	0.777	0.789	0.791	0.788	0.783
			0.596	0.715	0.807	0.874	0.918	0.945
		0.30	0.779	0.815	0.830	0.834	0.833	0.828
			0.593	0.712	0.805	0.873	0.920	0.948
		0.35	0.796	0.838	0.857	0.863	0.863	0.859
			0.589	0.708	0.801	0.869	0.916	0.946

Section modulus ratios: $\dfrac{Z_{inf}}{bh^2/6}$ and $\dfrac{Z_{sup}}{bh^2/6}$

$\dfrac{b_w}{b}$	$\dfrac{b_{inf}}{b}$	$\dfrac{h_{inf}}{h}$	$\dfrac{h_f}{h}=0.10$	$\dfrac{h_f}{h}=0.15$	$\dfrac{h_f}{h}=0.20$	$\dfrac{h_f}{h}=0.25$	$\dfrac{h_f}{h}=0.30$	$\dfrac{h_f}{h}=0.35$
0.20	0.90							
		0.10	0.551	0.565	0.569	0.567	0.564	0.561
			0.580	0.691	0.772	0.825	0.857	0.871
		0.15	0.652	0.671	0.677	0.677	0.673	0.669
			0.596	0.713	0.801	0.862	0.900	0.921
		0.20	0.727	0.753	0.762	0.763	0.759	0.755
			0.601	0.721	0.812	0.878	0.922	0.947
		0.25	0.780	0.812	0.825	0.827	0.825	0.820
			0.600	0.721	0.815	0.883	0.930	0.958
		0.30	0.813	0.851	0.868	0.873	0.872	0.867
			0.597	0.718	0.812	0.882	0.930	0.960
		0.35	0.830	0.875	0.896	0.903	0.903	0.899
			0.594	0.714	0.808	0.878	0.926	0.958
	0.95							
		0.10	0.571	0.586	0.589	0.588	0.585	0.581
			0.585	0.698	0.780	0.835	0.868	0.884
		0.15	0.678	0.699	0.705	0.704	0.701	0.696
			0.601	0.719	0.809	0.872	0.912	0.934
		0.20	0.758	0.785	0.795	0.796	0.792	0.787
			0.606	0.727	0.820	0.887	0.933	0.959
		0.25	0.812	0.847	0.861	0.864	0.861	0.857
			0.604	0.727	0.822	0.892	0.940	0.970
		0.30	0.846	0.887	0.906	0.912	0.911	0.906
			0.601	0.723	0.819	0.890	0.940	0.971
		0.35	0.863	0.911	0.934	0.942	0.943	0.939
			0.598	0.719	0.814	0.885	0.936	0.968
	1.00							
		0.10	0.590	0.606	0.610	0.608	0.605	0.601
			0.590	0.704	0.788	0.844	0.879	0.895
		0.15	0.704	0.726	0.732	0.732	0.728	0.724
			0.606	0.726	0.816	0.881	0.923	0.946
		0.20	0.788	0.816	0.827	0.828	0.825	0.820
			0.610	0.732	0.827	0.896	0.943	0.971
		0.25	0.844	0.881	0.896	0.900	0.898	0.893
			0.608	0.732	0.828	0.900	0.950	0.981
		0.30	0.879	0.923	0.943	0.950	0.949	0.944
			0.605	0.728	0.825	0.898	0.949	0.982
		0.35	0.895	0.946	0.971	0.981	0.982	0.978
			0.601	0.724	0.820	0.893	0.944	0.978

	Section modulus ratios	$\dfrac{Z_{inf}}{bh^2/6}$
		$\dfrac{Z_{sup}}{bh^2/6}$

$\dfrac{b_w}{b}$	$\dfrac{b_{inf}}{b}$	$\dfrac{h_{inf}}{h}$	$\dfrac{h_f}{h}=0.10$	$\dfrac{h_f}{h}=0.15$	$\dfrac{h_f}{h}=0.20$	$\dfrac{h_f}{h}=0.25$	$\dfrac{h_f}{h}=0.30$	$\dfrac{h_f}{h}=0.35$
0.25	0.25							
		0.10	0.324	0.333	0.335	0.335	0.334	0.333
			0.493	0.571	0.623	0.652	0.665	0.666
		0.15	0.324	0.333	0.335	0.335	0.334	0.333
			0.493	0.571	0.623	0.652	0.665	0.666
		0.20	0.324	0.333	0.335	0.335	0.334	0.333
			0.493	0.571	0.623	0.652	0.665	0.666
		0.25	0.324	0.333	0.335	0.335	0.334	0.333
			0.493	0.571	0.623	0.652	0.665	0.666
		0.30	0.324	0.333	0.335	0.335	0.334	0.333
			0.493	0.571	0.623	0.652	0.665	0.666
		0.35	0.324	0.333	0.335	0.335	0.334	0.333
			0.493	0.571	0.623	0.652	0.665	0.666
	0.30							
		0.10	0.344	0.353	0.356	0.355	0.354	0.354
			0.506	0.586	0.640	0.671	0.685	0.688
		0.15	0.351	0.361	0.364	0.363	0.362	0.361
			0.509	0.591	0.645	0.677	0.692	0.695
		0.20	0.357	0.368	0.371	0.370	0.369	0.368
			0.512	0.594	0.649	0.682	0.697	0.701
		0.25	0.362	0.373	0.376	0.376	0.374	0.373
			0.513	0.596	0.652	0.685	0.701	0.705
		0.30	0.366	0.377	0.380	0.380	0.379	0.378
			0.514	0.597	0.653	0.687	0.703	0.707
		0.35	0.369	0.380	0.384	0.383	0.382	0.381
			0.514	0.597	0.654	0.688	0.704	0.709
	0.35							
		0.10	0.363	0.373	0.376	0.376	0.374	0.374
			0.517	0.600	0.656	0.689	0.705	0.708
		0.15	0.378	0.389	0.392	0.392	0.390	0.389
			0.524	0.608	0.666	0.701	0.718	0.722
		0.20	0.391	0.402	0.406	0.405	0.404	0.403
			0.528	0.614	0.672	0.708	0.727	0.732
		0.25	0.401	0.413	0.416	0.416	0.414	0.413
			0.530	0.617	0.677	0.714	0.733	0.739
		0.30	0.408	0.421	0.425	0.424	0.423	0.422
			0.531	0.618	0.679	0.716	0.736	0.743
		0.35	0.414	0.427	0.431	0.431	0.429	0.428
			0.531	0.618	0.679	0.718	0.738	0.745

Section modulus ratios $\dfrac{Z_{inf}}{bh^2/6}$ $\dfrac{Z_{sup}}{bh^2/6}$

$\dfrac{b_w}{b}$	$\dfrac{b_{inf}}{b}$	$\dfrac{h_{inf}}{h}$	$\dfrac{h_f}{h}=0.10$	$\dfrac{h_f}{h}=0.15$	$\dfrac{h_f}{h}=0.20$	$\dfrac{h_f}{h}=0.25$	$\dfrac{h_f}{h}=0.30$	$\dfrac{h_f}{h}=0.35$
0.25	0.40							
		0.10	0.383	0.394	0.397	0.396	0.395	0.394
			0.528	0.613	0.671	0.706	0.723	0.728
		0.15	0.406	0.417	0.421	0.420	0.418	0.417
			0.537	0.624	0.684	0.722	0.741	0.747
		0.20	0.424	0.436	0.440	0.440	0.438	0.437
			0.542	0.631	0.693	0.732	0.753	0.760
		0.25	0.438	0.452	0.456	0.456	0.454	0.453
			0.545	0.635	0.698	0.739	0.761	0.769
		0.30	0.449	0.463	0.468	0.468	0.467	0.465
			0.546	0.636	0.701	0.742	0.765	0.774
		0.35	0.457	0.472	0.477	0.478	0.476	0.474
			0.545	0.636	0.701	0.743	0.767	0.777
	0.45							
		0.10	0.403	0.414	0.417	0.417	0.415	0.414
			0.538	0.625	0.685	0.722	0.741	0.746
		0.15	0.433	0.445	0.449	0.448	0.446	0.445
			0.549	0.639	0.702	0.741	0.762	0.770
		0.20	0.457	0.470	0.475	0.474	0.473	0.471
			0.555	0.647	0.712	0.754	0.777	0.786
		0.25	0.475	0.490	0.495	0.495	0.493	0.492
			0.558	0.651	0.718	0.761	0.786	0.796
		0.30	0.489	0.505	0.511	0.512	0.510	0.508
			0.559	0.652	0.720	0.765	0.790	0.802
		0.35	0.499	0.516	0.523	0.524	0.522	0.520
			0.558	0.652	0.720	0.765	0.792	0.804
	0.50							
		0.10	0.422	0.434	0.438	0.437	0.435	0.434
			0.547	0.636	0.698	0.737	0.757	0.764
		0.15	0.459	0.473	0.477	0.476	0.474	0.473
			0.560	0.652	0.717	0.759	0.783	0.792
		0.20	0.489	0.504	0.509	0.509	0.507	0.505
			0.566	0.661	0.729	0.773	0.799	0.810
		0.25	0.512	0.528	0.534	0.535	0.532	0.530
			0.569	0.665	0.735	0.781	0.808	0.821
		0.30	0.528	0.547	0.554	0.554	0.552	0.550
			0.570	0.666	0.737	0.785	0.813	0.827
		0.35	0.540	0.560	0.568	0.569	0.567	0.565
			0.568	0.665	0.736	0.785	0.814	0.829

Section modulus ratios

$$\frac{Z_{inf}}{bh^2/6}$$

$$\frac{Z_{sup}}{bh^2/6}$$

$\dfrac{b_w}{b}$	$\dfrac{b_{inf}}{b}$	$\dfrac{h_{inf}}{h}$	$\dfrac{h_f}{h}=0.10$	$\dfrac{h_f}{h}=0.15$	$\dfrac{h_f}{h}=0.20$	$\dfrac{h_f}{h}=0.25$	$\dfrac{h_f}{h}=0.30$	$\dfrac{h_f}{h}=0.35$
0.25	0.55							
		0.10	0.442	0.454	0.458	0.457	0.455	0.454
			0.556	0.647	0.711	0.751	0.773	0.781
		0.15	0.486	0.500	0.505	0.504	0.502	0.500
			0.570	0.664	0.732	0.776	0.801	0.812
		0.20	0.521	0.537	0.543	0.543	0.541	0.538
			0.577	0.674	0.744	0.791	0.819	0.832
		0.25	0.548	0.566	0.573	0.573	0.571	0.569
			0.580	0.678	0.750	0.799	0.829	0.844
		0.30	0.567	0.587	0.595	0.596	0.595	0.592
			0.580	0.679	0.752	0.802	0.833	0.849
		0.35	0.580	0.602	0.611	0.613	0.612	0.609
			0.578	0.677	0.751	0.802	0.834	0.851
	0.60							
		0.10	0.461	0.474	0.478	0.478	0.476	0.474
			0.564	0.657	0.722	0.764	0.788	0.797
		0.15	0.513	0.528	0.533	0.532	0.530	0.528
			0.579	0.676	0.745	0.792	0.819	0.831
		0.20	0.553	0.570	0.577	0.577	0.574	0.572
			0.586	0.685	0.758	0.808	0.837	0.852
		0.25	0.583	0.603	0.611	0.612	0.610	0.607
			0.589	0.689	0.764	0.816	0.848	0.864
		0.30	0.605	0.627	0.636	0.638	0.636	0.633
			0.588	0.690	0.765	0.818	0.852	0.870
		0.35	0.619	0.644	0.654	0.657	0.655	0.653
			0.586	0.688	0.764	0.818	0.852	0.871
	0.65							
		0.10	0.481	0.494	0.498	0.498	0.496	0.494
			0.572	0.666	0.734	0.777	0.802	0.812
		0.15	0.539	0.555	0.560	0.560	0.558	0.555
			0.587	0.686	0.758	0.806	0.835	0.849
		0.20	0.585	0.603	0.610	0.610	0.608	0.605
			0.595	0.696	0.771	0.823	0.855	0.871
		0.25	0.619	0.640	0.648	0.650	0.648	0.644
			0.597	0.700	0.777	0.830	0.865	0.883
		0.30	0.642	0.667	0.677	0.679	0.677	0.674
			0.596	0.700	0.778	0.833	0.868	0.888
		0.35	0.657	0.685	0.697	0.700	0.698	0.695
			0.594	0.697	0.776	0.831	0.868	0.889

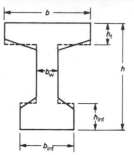

	Section modulus ratios	$\dfrac{Z_{inf}}{bh^2/6}$
		$\dfrac{Z_{sup}}{bh^2/6}$

$\dfrac{b_w}{b}$	$\dfrac{b_{inf}}{b}$	$\dfrac{h_{inf}}{h}$	$\dfrac{h_f}{h}=0.10$	$\dfrac{h_f}{h}=0.15$	$\dfrac{h_f}{h}=0.20$	$\dfrac{h_f}{h}=0.25$	$\dfrac{h_f}{h}=0.30$	$\dfrac{h_f}{h}=0.35$
0.25	0.70							
		0.10	0.500	0.514	0.519	0.518	0.516	0.514
			0.579	0.675	0.744	0.789	0.815	0.826
		0.15	0.565	0.582	0.588	0.588	0.585	0.583
			0.595	0.696	0.770	0.820	0.850	0.865
		0.20	0.616	0.636	0.643	0.644	0.642	0.638
			0.603	0.706	0.783	0.837	0.870	0.888
		0.25	0.653	0.676	0.686	0.687	0.685	0.682
			0.605	0.709	0.788	0.844	0.880	0.900
		0.30	0.679	0.705	0.717	0.720	0.718	0.715
			0.603	0.709	0.789	0.846	0.883	0.905
		0.35	0.695	0.724	0.738	0.742	0.741	0.738
			0.601	0.706	0.786	0.844	0.882	0.905
	0.75							
		0.10	0.520	0.534	0.539	0.538	0.536	0.534
			0.586	0.684	0.754	0.801	0.828	0.840
		0.15	0.592	0.609	0.615	0.615	0.613	0.610
			0.603	0.705	0.781	0.833	0.865	0.881
		0.20	0.647	0.668	0.676	0.677	0.675	0.671
			0.610	0.715	0.794	0.849	0.885	0.904
		0.25	0.687	0.712	0.723	0.725	0.723	0.719
			0.611	0.718	0.799	0.857	0.894	0.916
		0.30	0.715	0.743	0.756	0.760	0.758	0.755
			0.610	0.717	0.799	0.858	0.897	0.921
		0.35	0.731	0.764	0.779	0.783	0.783	0.780
			0.607	0.714	0.796	0.855	0.896	0.920
	0.80							
		0.10	0.539	0.554	0.559	0.558	0.556	0.553
			0.593	0.692	0.764	0.812	0.840	0.854
		0.15	0.618	0.636	0.643	0.643	0.640	0.637
			0.610	0.714	0.791	0.845	0.878	0.896
		0.20	0.678	0.700	0.709	0.710	0.708	0.704
			0.617	0.723	0.804	0.861	0.899	0.920
		0.25	0.721	0.748	0.759	0.762	0.760	0.756
			0.618	0.726	0.808	0.868	0.908	0.931
		0.30	0.750	0.781	0.795	0.799	0.798	0.794
			0.616	0.724	0.808	0.869	0.910	0.935
		0.35	0.767	0.802	0.819	0.824	0.824	0.821
			0.613	0.721	0.804	0.866	0.908	0.934

Section modulus ratios $\dfrac{Z_{inf}}{bh^2/6}$ $\dfrac{Z_{sup}}{bh^2/6}$

$\dfrac{b_w}{b}$	$\dfrac{b_{inf}}{b}$	$\dfrac{h_{inf}}{h}$	$\dfrac{h_f}{h}=0.10$	$\dfrac{h_f}{h}=0.15$	$\dfrac{h_f}{h}=0.20$	$\dfrac{h_f}{h}=0.25$	$\dfrac{h_f}{h}=0.30$	$\dfrac{h_f}{h}=0.35$
0.25	0.85							
		0.10	0.558	0.574	0.579	0.579	0.576	0.573
			0.599	0.699	0.773	0.822	0.852	0.867
		0.15	0.643	0.663	0.670	0.670	0.668	0.664
			0.616	0.722	0.801	0.856	0.891	0.910
		0.20	0.708	0.732	0.742	0.743	0.741	0.737
			0.623	0.731	0.813	0.872	0.911	0.934
		0.25	0.754	0.783	0.795	0.798	0.796	0.793
			0.624	0.733	0.817	0.879	0.920	0.945
		0.30	0.784	0.818	0.833	0.838	0.837	0.833
			0.621	0.731	0.816	0.879	0.922	0.948
		0.35	0.802	0.839	0.858	0.864	0.864	0.861
			0.618	0.728	0.813	0.876	0.919	0.947
	0.90							
		0.10	0.578	0.594	0.599	0.599	0.596	0.593
			0.605	0.707	0.782	0.833	0.863	0.879
		0.15	0.669	0.690	0.697	0.698	0.695	0.691
			0.622	0.729	0.810	0.866	0.903	0.923
		0.20	0.738	0.764	0.774	0.776	0.773	0.769
			0.629	0.738	0.822	0.883	0.923	0.947
		0.25	0.787	0.818	0.831	0.835	0.833	0.829
			0.629	0.740	0.826	0.889	0.931	0.958
		0.30	0.818	0.854	0.871	0.877	0.876	0.872
			0.627	0.738	0.824	0.888	0.932	0.960
		0.35	0.836	0.876	0.896	0.904	0.904	0.901
			0.623	0.734	0.820	0.884	0.930	0.958
	0.95							
		0.10	0.597	0.614	0.619	0.619	0.616	0.613
			0.611	0.714	0.790	0.842	0.874	0.891
		0.15	0.695	0.716	0.724	0.725	0.722	0.718
			0.628	0.736	0.818	0.876	0.914	0.936
		0.20	0.768	0.795	0.806	0.808	0.806	0.801
			0.634	0.745	0.830	0.893	0.934	0.960
		0.25	0.819	0.852	0.867	0.871	0.869	0.865
			0.634	0.746	0.833	0.898	0.942	0.970
		0.30	0.852	0.890	0.908	0.915	0.914	0.910
			0.631	0.743	0.831	0.897	0.943	0.972
		0.35	0.869	0.913	0.934	0.943	0.944	0.941
			0.628	0.739	0.827	0.893	0.939	0.969

Section modulus ratios
$$\frac{Z_{inf}}{bh^2/6}$$
$$\frac{Z_{sup}}{bh^2/6}$$

$\dfrac{b_w}{b}$	$\dfrac{b_{inf}}{b}$	$\dfrac{h_{inf}}{h}$	$\dfrac{h_f}{h}=0.10$	$\dfrac{h_f}{h}=0.15$	$\dfrac{h_f}{h}=0.20$	$\dfrac{h_f}{h}=0.25$	$\dfrac{h_f}{h}=0.30$	$\dfrac{h_f}{h}=0.35$
0.25	1.00							
		0.10	0.616	0.633	0.639	0.639	0.636	0.633
			0.616	0.720	0.798	0.851	0.885	0.902
		0.15	0.720	0.743	0.751	0.752	0.749	0.745
			0.633	0.743	0.826	0.886	0.925	0.948
		0.20	0.798	0.826	0.838	0.840	0.838	0.834
			0.639	0.751	0.838	0.902	0.945	0.972
		0.25	0.851	0.886	0.902	0.906	0.905	0.901
			0.639	0.752	0.840	0.906	0.952	0.981
		0.30	0.885	0.925	0.945	0.952	0.952	0.948
			0.636	0.749	0.838	0.905	0.952	0.983
		0.35	0.902	0.948	0.972	0.981	0.983	0.980
			0.633	0.745	0.834	0.901	0.948	0.980
0.30	0.30							
		0.10	0.375	0.386	0.390	0.390	0.389	0.388
			0.529	0.604	0.656	0.687	0.703	0.708
		0.15	0.375	0.386	0.390	0.390	0.389	0.388
			0.529	0.604	0.656	0.687	0.703	0.708
		0.20	0.375	0.386	0.390	0.390	0.389	0.388
			0.529	0.604	0.656	0.687	0.703	0.708
		0.25	0.375	0.386	0.390	0.390	0.389	0.388
			0.529	0.604	0.656	0.687	0.703	0.708
		0.30	0.375	0.386	0.390	0.390	0.389	0.388
			0.529	0.604	0.656	0.687	0.703	0.708
		0.35	0.375	0.386	0.390	0.390	0.389	0.388
			0.529	0.604	0.656	0.687	0.703	0.708
	0.35							
		0.10	0.394	0.406	0.410	0.410	0.408	0.407
			0.540	0.617	0.671	0.704	0.721	0.727
		0.15	0.402	0.413	0.417	0.418	0.416	0.415
			0.543	0.621	0.675	0.709	0.727	0.733
		0.20	0.408	0.420	0.424	0.424	0.423	0.422
			0.545	0.624	0.679	0.713	0.731	0.737
		0.25	0.412	0.425	0.429	0.429	0.428	0.427
			0.547	0.626	0.681	0.716	0.734	0.741
		0.30	0.416	0.429	0.433	0.434	0.432	0.431
			0.547	0.627	0.682	0.717	0.736	0.743
		0.35	0.419	0.432	0.436	0.437	0.435	0.434
			0.547	0.627	0.682	0.718	0.737	0.744

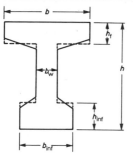

Section modulus ratios	$\dfrac{Z_{inf}}{bh^2/6}$
	$\dfrac{Z_{sup}}{bh^2/6}$

$\dfrac{b_w}{b}$	$\dfrac{b_{inf}}{b}$	$\dfrac{h_{inf}}{h}$	$\dfrac{h_f}{h}=0.10$	$\dfrac{h_f}{h}=0.15$	$\dfrac{h_f}{h}=0.20$	$\dfrac{h_f}{h}=0.25$	$\dfrac{h_f}{h}=0.30$	$\dfrac{h_f}{h}=0.35$
0.30	0.40							
		0.10	0.413	0.426	0.430	0.430	0.428	0.427
			0.550	0.630	0.685	0.720	0.738	0.744
		0.15	0.428	0.441	0.445	0.445	0.444	0.443
			0.557	0.637	0.694	0.730	0.749	0.756
		0.20	0.440	0.453	0.458	0.458	0.457	0.455
			0.561	0.642	0.699	0.736	0.757	0.765
		0.25	0.449	0.463	0.468	0.469	0.467	0.466
			0.563	0.645	0.703	0.741	0.762	0.770
		0.30	0.456	0.471	0.476	0.477	0.475	0.474
			0.564	0.646	0.705	0.743	0.765	0.774
		0.35	0.462	0.476	0.482	0.483	0.481	0.480
			0.563	0.646	0.705	0.744	0.766	0.775
	0.45							
		0.10	0.433	0.445	0.450	0.450	0.448	0.447
			0.560	0.641	0.698	0.734	0.754	0.762
		0.15	0.455	0.468	0.473	0.473	0.472	0.470
			0.569	0.652	0.710	0.748	0.770	0.778
		0.20	0.472	0.486	0.492	0.492	0.491	0.489
			0.574	0.658	0.718	0.758	0.780	0.790
		0.25	0.486	0.501	0.507	0.508	0.506	0.504
			0.577	0.662	0.723	0.763	0.787	0.797
		0.30	0.496	0.512	0.518	0.519	0.518	0.516
			0.578	0.663	0.725	0.766	0.790	0.801
		0.35	0.503	0.520	0.527	0.528	0.527	0.525
			0.577	0.663	0.725	0.767	0.791	0.803
	0.50							
		0.10	0.452	0.465	0.470	0.470	0.468	0.467
			0.570	0.652	0.711	0.749	0.769	0.778
		0.15	0.481	0.495	0.500	0.501	0.499	0.497
			0.580	0.665	0.726	0.766	0.789	0.799
		0.20	0.504	0.519	0.525	0.526	0.524	0.522
			0.587	0.673	0.735	0.777	0.801	0.813
		0.25	0.522	0.539	0.545	0.546	0.544	0.542
			0.589	0.677	0.741	0.783	0.809	0.821
		0.30	0.535	0.553	0.560	0.562	0.560	0.558
			0.590	0.678	0.743	0.786	0.813	0.826
		0.35	0.544	0.563	0.571	0.573	0.572	0.569
			0.589	0.678	0.743	0.787	0.814	0.828

Section modulus ratios $\dfrac{Z_{inf}}{bh^2/6}$ $\dfrac{Z_{sup}}{bh^2/6}$

$\dfrac{b_w}{b}$	$\dfrac{b_{inf}}{b}$	$\dfrac{h_{inf}}{h}$	$\dfrac{h_f}{h}=0.10$	$\dfrac{h_f}{h}=0.15$	$\dfrac{h_f}{h}=0.20$	$\dfrac{h_f}{h}=0.25$	$\dfrac{h_f}{h}=0.30$	$\dfrac{h_f}{h}=0.35$
0.30	0.55							
		0.10	0.471	0.485	0.490	0.490	0.488	0.486
			0.579	0.663	0.723	0.762	0.784	0.793
		0.15	0.507	0.522	0.528	0.528	0.526	0.524
			0.591	0.678	0.740	0.782	0.807	0.818
		0.20	0.536	0.552	0.559	0.559	0.558	0.556
			0.598	0.686	0.751	0.795	0.821	0.834
		0.25	0.558	0.576	0.583	0.584	0.583	0.580
			0.601	0.690	0.756	0.802	0.830	0.844
		0.30	0.574	0.593	0.601	0.603	0.602	0.599
			0.601	0.691	0.758	0.805	0.833	0.849
		0.35	0.584	0.605	0.615	0.617	0.616	0.613
			0.600	0.691	0.758	0.805	0.834	0.850
	0.60							
		0.10	0.490	0.504	0.509	0.510	0.508	0.506
			0.587	0.673	0.734	0.775	0.798	0.808
		0.15	0.533	0.549	0.555	0.556	0.554	0.552
			0.601	0.689	0.754	0.798	0.824	0.836
		0.20	0.567	0.585	0.592	0.593	0.591	0.589
			0.608	0.699	0.765	0.811	0.839	0.854
		0.25	0.593	0.612	0.620	0.622	0.620	0.618
			0.611	0.703	0.771	0.818	0.848	0.864
		0.30	0.611	0.633	0.642	0.644	0.643	0.640
			0.611	0.703	0.773	0.821	0.852	0.869
		0.35	0.623	0.647	0.657	0.660	0.659	0.656
			0.610	0.702	0.772	0.821	0.852	0.870
	0.65							
		0.10	0.509	0.524	0.529	0.530	0.528	0.526
			0.595	0.682	0.745	0.787	0.812	0.823
		0.15	0.559	0.576	0.582	0.583	0.581	0.579
			0.610	0.700	0.767	0.812	0.839	0.853
		0.20	0.598	0.617	0.625	0.626	0.624	0.621
			0.617	0.710	0.779	0.826	0.856	0.872
		0.25	0.627	0.649	0.657	0.659	0.658	0.655
			0.620	0.714	0.784	0.834	0.865	0.883
		0.30	0.648	0.672	0.682	0.685	0.683	0.681
			0.620	0.714	0.786	0.836	0.869	0.887
		0.35	0.662	0.687	0.699	0.702	0.702	0.699
			0.619	0.713	0.784	0.835	0.869	0.888

| | Section modulus ratios | $\dfrac{Z_{inf}}{bh^2/6}$ $\dfrac{Z_{sup}}{bh^2/6}$ |

$\dfrac{b_w}{b}$	$\dfrac{b_{inf}}{b}$	$\dfrac{h_{inf}}{h}$	$\dfrac{h_f}{h}=0.10$	$\dfrac{h_f}{h}=0.15$	$\dfrac{h_f}{h}=0.20$	$\dfrac{h_f}{h}=0.25$	$\dfrac{h_f}{h}=0.30$	$\dfrac{h_f}{h}=0.35$
0.30	0.70							
		0.10	0.528	0.543	0.549	0.549	0.547	0.545
			0.603	0.691	0.756	0.799	0.824	0.836
		0.15	0.585	0.602	0.609	0.610	0.608	0.606
			0.618	0.710	0.778	0.825	0.854	0.869
		0.20	0.629	0.649	0.657	0.659	0.657	0.654
			0.626	0.720	0.791	0.840	0.872	0.889
		0.25	0.662	0.684	0.694	0.697	0.695	0.692
			0.629	0.724	0.796	0.848	0.881	0.900
		0.30	0.685	0.710	0.721	0.725	0.724	0.721
			0.628	0.724	0.797	0.850	0.884	0.905
		0.35	0.699	0.727	0.740	0.744	0.744	0.741
			0.626	0.722	0.796	0.848	0.884	0.905
	0.75							
		0.10	0.547	0.563	0.569	0.569	0.567	0.565
			0.610	0.700	0.765	0.810	0.837	0.850
		0.15	0.611	0.629	0.636	0.637	0.635	0.632
			0.626	0.720	0.790	0.838	0.868	0.885
		0.20	0.660	0.681	0.690	0.691	0.690	0.687
			0.634	0.730	0.802	0.854	0.887	0.905
		0.25	0.696	0.720	0.731	0.733	0.732	0.729
			0.637	0.734	0.808	0.861	0.896	0.916
		0.30	0.720	0.748	0.760	0.764	0.763	0.760
			0.636	0.733	0.808	0.862	0.899	0.920
		0.35	0.736	0.766	0.780	0.785	0.785	0.782
			0.634	0.731	0.806	0.861	0.898	0.920
	0.80							
		0.10	0.566	0.582	0.588	0.589	0.587	0.585
			0.617	0.708	0.775	0.821	0.848	0.862
		0.15	0.636	0.655	0.663	0.664	0.662	0.659
			0.634	0.729	0.800	0.850	0.882	0.899
		0.20	0.690	0.712	0.722	0.724	0.722	0.719
			0.642	0.739	0.813	0.866	0.900	0.920
		0.25	0.729	0.755	0.767	0.770	0.769	0.765
			0.644	0.742	0.818	0.873	0.909	0.931
		0.30	0.756	0.785	0.799	0.803	0.803	0.799
			0.643	0.742	0.818	0.874	0.912	0.935
		0.35	0.772	0.804	0.820	0.826	0.826	0.823
			0.640	0.739	0.815	0.872	0.910	0.934

Section modulus ratios
$$\frac{Z_{inf}}{bh^2/6}$$
$$\frac{Z_{sup}}{bh^2/6}$$

$\dfrac{b_w}{b}$	$\dfrac{b_{inf}}{b}$	$\dfrac{h_{inf}}{h}$	$\dfrac{h_f}{h}=0.10$	$\dfrac{h_f}{h}=0.15$	$\dfrac{h_f}{h}=0.20$	$\dfrac{h_f}{h}=0.25$	$\dfrac{h_f}{h}=0.30$	$\dfrac{h_f}{h}=0.35$
0.30	0.85							
		0.10	0.585	0.602	0.608	0.609	0.607	0.604
			0.624	0.716	0.784	0.831	0.860	0.874
		0.15	0.662	0.682	0.690	0.691	0.689	0.686
			0.641	0.737	0.810	0.861	0.894	0.913
		0.20	0.720	0.744	0.754	0.756	0.755	0.751
			0.649	0.747	0.823	0.877	0.913	0.935
		0.25	0.762	0.790	0.802	0.806	0.805	0.801
			0.650	0.750	0.827	0.884	0.922	0.945
		0.30	0.790	0.822	0.837	0.842	0.841	0.838
			0.649	0.749	0.827	0.884	0.924	0.948
		0.35	0.807	0.842	0.859	0.866	0.866	0.863
			0.646	0.746	0.824	0.882	0.922	0.947
	0.90							
		0.10	0.604	0.621	0.628	0.628	0.626	0.624
			0.630	0.723	0.793	0.841	0.871	0.886
		0.15	0.687	0.708	0.716	0.718	0.716	0.712
			0.648	0.745	0.819	0.872	0.906	0.926
		0.20	0.750	0.775	0.786	0.788	0.787	0.783
			0.655	0.755	0.832	0.888	0.925	0.948
		0.25	0.795	0.824	0.838	0.842	0.841	0.837
			0.657	0.758	0.836	0.894	0.933	0.958
		0.30	0.824	0.858	0.874	0.880	0.880	0.877
			0.655	0.756	0.835	0.894	0.935	0.961
		0.35	0.841	0.878	0.897	0.905	0.906	0.903
			0.652	0.753	0.832	0.891	0.933	0.959
	0.95							
		0.10	0.623	0.641	0.647	0.648	0.646	0.643
			0.636	0.730	0.801	0.850	0.881	0.897
		0.15	0.712	0.734	0.743	0.745	0.742	0.739
			0.654	0.753	0.828	0.882	0.918	0.938
		0.20	0.779	0.806	0.817	0.820	0.819	0.815
			0.661	0.763	0.841	0.898	0.937	0.960
		0.25	0.827	0.858	0.873	0.877	0.876	0.873
			0.662	0.765	0.844	0.903	0.944	0.970
		0.30	0.858	0.893	0.911	0.918	0.918	0.915
			0.660	0.763	0.843	0.903	0.945	0.973
		0.35	0.875	0.915	0.935	0.944	0.945	0.942
			0.657	0.759	0.840	0.900	0.943	0.971

Section modulus ratios

$$\frac{Z_{inf}}{bh^2/6}$$

$$\frac{Z_{sup}}{bh^2/6}$$

$\dfrac{b_w}{b}$	$\dfrac{b_{inf}}{b}$	$\dfrac{h_{inf}}{h}$	$\dfrac{h_f}{h}=0.10$	$\dfrac{h_f}{h}=0.15$	$\dfrac{h_f}{h}=0.20$	$\dfrac{h_f}{h}=0.25$	$\dfrac{h_f}{h}=0.30$	$\dfrac{h_f}{h}=0.35$
0.30	1.00							
		0.10	0.642	0.660	0.667	0.668	0.666	0.662
			0.642	0.737	0.809	0.859	0.891	0.908
		0.15	0.737	0.760	0.769	0.771	0.769	0.765
			0.660	0.760	0.836	0.892	0.928	0.950
		0.20	0.809	0.836	0.849	0.852	0.851	0.847
			0.667	0.769	0.849	0.907	0.947	0.972
		0.25	0.859	0.892	0.907	0.913	0.912	0.908
			0.668	0.771	0.852	0.913	0.955	0.982
		0.30	0.891	0.928	0.947	0.955	0.955	0.952
			0.666	0.769	0.851	0.912	0.955	0.984
		0.35	0.908	0.950	0.972	0.982	0.984	0.981
			0.662	0.765	0.847	0.908	0.952	0.981
0.35	0.35							
		0.10	0.424	0.437	0.442	0.442	0.441	0.440
			0.563	0.635	0.686	0.718	0.736	0.743
		0.15	0.424	0.437	0.442	0.442	0.441	0.440
			0.563	0.635	0.686	0.718	0.736	0.743
		0.20	0.424	0.437	0.442	0.442	0.441	0.440
			0.563	0.635	0.686	0.718	0.736	0.743
		0.25	0.424	0.437	0.442	0.442	0.441	0.440
			0.563	0.635	0.686	0.718	0.736	0.743
		0.30	0.424	0.437	0.442	0.442	0.441	0.440
			0.563	0.635	0.686	0.718	0.736	0.743
		0.35	0.424	0.437	0.442	0.442	0.441	0.440
			0.563	0.635	0.686	0.718	0.736	0.743
	0.40							
		0.10	0.443	0.456	0.461	0.462	0.461	0.459
			0.574	0.647	0.699	0.733	0.752	0.759
		0.15	0.450	0.463	0.469	0.470	0.468	0.467
			0.577	0.651	0.703	0.738	0.757	0.765
		0.20	0.456	0.470	0.475	0.476	0.475	0.473
			0.579	0.653	0.706	0.741	0.761	0.769
		0.25	0.461	0.474	0.480	0.481	0.480	0.478
			0.580	0.655	0.708	0.743	0.763	0.772
		0.30	0.464	0.478	0.484	0.485	0.484	0.482
			0.581	0.656	0.709	0.744	0.765	0.773
		0.35	0.467	0.481	0.487	0.488	0.487	0.485
			0.581	0.656	0.709	0.745	0.765	0.774

Section modulus ratios
$$\frac{Z_{inf}}{bh^2/6}$$
$$\frac{Z_{sup}}{bh^2/6}$$

$\frac{b_w}{b}$	$\frac{b_{inf}}{b}$	$\frac{h_{inf}}{h}$	$\frac{h_f}{h}=0.10$	$\frac{h_f}{h}=0.15$	$\frac{h_f}{h}=0.20$	$\frac{h_f}{h}=0.25$	$\frac{h_f}{h}=0.30$	$\frac{h_f}{h}=0.35$
0.35	0.45							
		0.10	0.462	0.475	0.481	0.482	0.480	0.479
			0.584	0.658	0.712	0.747	0.767	0.775
		0.15	0.476	0.490	0.496	0.497	0.496	0.494
			0.590	0.665	0.720	0.756	0.777	0.786
		0.20	0.488	0.502	0.508	0.509	0.508	0.506
			0.593	0.670	0.725	0.762	0.783	0.793
		0.25	0.497	0.512	0.518	0.519	0.518	0.516
			0.595	0.672	0.728	0.766	0.788	0.798
		0.30	0.503	0.519	0.526	0.527	0.526	0.524
			0.596	0.674	0.730	0.768	0.790	0.801
		0.35	0.508	0.524	0.531	0.533	0.532	0.530
			0.596	0.674	0.730	0.768	0.791	0.802
	0.50							
		0.10	0.481	0.495	0.500	0.501	0.500	0.498
			0.593	0.669	0.724	0.760	0.781	0.791
		0.15	0.502	0.517	0.523	0.524	0.523	0.521
			0.601	0.679	0.735	0.773	0.795	0.805
		0.20	0.519	0.535	0.541	0.542	0.541	0.539
			0.606	0.685	0.742	0.781	0.804	0.816
		0.25	0.532	0.549	0.556	0.557	0.556	0.554
			0.609	0.688	0.747	0.786	0.810	0.822
		0.30	0.542	0.559	0.567	0.568	0.567	0.565
			0.610	0.690	0.748	0.788	0.813	0.825
		0.35	0.549	0.567	0.575	0.577	0.576	0.574
			0.610	0.689	0.748	0.789	0.814	0.827
	0.55							
		0.10	0.500	0.514	0.520	0.521	0.519	0.518
			0.602	0.680	0.736	0.773	0.795	0.805
		0.15	0.528	0.543	0.550	0.551	0.550	0.548
			0.612	0.692	0.750	0.789	0.812	0.824
		0.20	0.550	0.567	0.574	0.575	0.574	0.572
			0.618	0.699	0.758	0.799	0.824	0.836
		0.25	0.568	0.585	0.593	0.595	0.594	0.591
			0.621	0.703	0.763	0.805	0.830	0.844
		0.30	0.580	0.599	0.607	0.610	0.608	0.606
			0.622	0.704	0.765	0.807	0.834	0.848
		0.35	0.589	0.609	0.618	0.620	0.619	0.617
			0.622	0.704	0.765	0.808	0.835	0.849

Section modulus ratios:
$$\frac{Z_{inf}}{bh^2/6}$$
$$\frac{Z_{sup}}{bh^2/6}$$

$\dfrac{b_w}{b}$	$\dfrac{b_{inf}}{b}$	$\dfrac{h_{inf}}{h}$	$\dfrac{h_f}{h}=0.10$	$\dfrac{h_f}{h}=0.15$	$\dfrac{h_f}{h}=0.20$	$\dfrac{h_f}{h}=0.25$	$\dfrac{h_f}{h}=0.30$	$\dfrac{h_f}{h}=0.35$
0.35	0.60							
		0.10	0.518	0.533	0.539	0.540	0.539	0.537
			0.611	0.689	0.747	0.785	0.809	0.819
		0.15	0.553	0.570	0.576	0.578	0.576	0.574
			0.623	0.704	0.763	0.804	0.829	0.841
		0.20	0.581	0.599	0.606	0.608	0.607	0.605
			0.629	0.712	0.773	0.815	0.842	0.856
		0.25	0.602	0.621	0.630	0.632	0.631	0.629
			0.633	0.716	0.778	0.822	0.849	0.864
		0.30	0.618	0.638	0.647	0.650	0.649	0.647
			0.633	0.717	0.780	0.824	0.853	0.868
		0.35	0.628	0.650	0.660	0.663	0.662	0.660
			0.632	0.716	0.779	0.824	0.853	0.869
	0.65							
		0.10	0.537	0.552	0.559	0.560	0.558	0.557
			0.619	0.699	0.757	0.797	0.821	0.833
		0.15	0.579	0.596	0.603	0.605	0.603	0.601
			0.632	0.715	0.776	0.818	0.844	0.858
		0.20	0.612	0.631	0.639	0.641	0.639	0.637
			0.640	0.724	0.787	0.831	0.859	0.874
		0.25	0.637	0.657	0.666	0.669	0.668	0.665
			0.643	0.728	0.792	0.837	0.866	0.883
		0.30	0.655	0.677	0.687	0.690	0.689	0.687
			0.643	0.729	0.793	0.840	0.870	0.887
		0.35	0.666	0.690	0.701	0.705	0.705	0.702
			0.642	0.728	0.793	0.839	0.870	0.888
	0.70							
		0.10	0.556	0.572	0.578	0.579	0.578	0.576
			0.627	0.708	0.768	0.808	0.833	0.846
		0.15	0.604	0.622	0.630	0.631	0.630	0.628
			0.641	0.725	0.788	0.831	0.859	0.874
		0.20	0.642	0.662	0.671	0.673	0.672	0.669
			0.649	0.735	0.799	0.845	0.874	0.891
		0.25	0.671	0.693	0.703	0.706	0.705	0.702
			0.652	0.739	0.805	0.852	0.882	0.900
		0.30	0.691	0.715	0.726	0.730	0.729	0.727
			0.652	0.739	0.806	0.854	0.886	0.904
		0.35	0.704	0.730	0.742	0.747	0.746	0.744
			0.651	0.738	0.805	0.853	0.885	0.905

Section modulus ratios
$$\frac{Z_{inf}}{bh^2/6} \qquad \frac{Z_{sup}}{bh^2/6}$$

$\dfrac{b_w}{b}$	$\dfrac{b_{inf}}{b}$	$\dfrac{h_{inf}}{h}$	$\dfrac{h_f}{h}=0.10$	$\dfrac{h_f}{h}=0.15$	$\dfrac{h_f}{h}=0.20$	$\dfrac{h_f}{h}=0.25$	$\dfrac{h_f}{h}=0.30$	$\dfrac{h_f}{h}=0.35$
0.35	0.75							
		0.10	0.574	0.591	0.598	0.599	0.597	0.595
			0.634	0.716	0.777	0.819	0.845	0.859
		0.15	0.630	0.648	0.656	0.658	0.657	0.654
			0.650	0.735	0.799	0.844	0.873	0.888
		0.20	0.673	0.693	0.703	0.705	0.704	0.701
			0.658	0.745	0.811	0.858	0.889	0.907
		0.25	0.705	0.728	0.739	0.742	0.741	0.738
			0.661	0.749	0.816	0.865	0.897	0.916
		0.30	0.727	0.752	0.765	0.769	0.768	0.766
			0.661	0.749	0.817	0.867	0.900	0.920
		0.35	0.741	0.768	0.782	0.787	0.787	0.785
			0.659	0.747	0.816	0.866	0.899	0.920
	0.80							
		0.10	0.593	0.610	0.617	0.618	0.617	0.614
			0.641	0.725	0.787	0.830	0.857	0.871
		0.15	0.655	0.674	0.683	0.685	0.683	0.680
			0.658	0.744	0.810	0.856	0.886	0.903
		0.20	0.703	0.725	0.735	0.737	0.736	0.733
			0.666	0.755	0.822	0.870	0.903	0.921
		0.25	0.738	0.762	0.774	0.778	0.777	0.774
			0.669	0.758	0.827	0.877	0.911	0.931
		0.30	0.762	0.789	0.803	0.807	0.807	0.805
			0.668	0.758	0.828	0.879	0.913	0.935
		0.35	0.777	0.807	0.822	0.828	0.828	0.826
			0.666	0.756	0.826	0.877	0.912	0.935
	0.85							
		0.10	0.612	0.629	0.636	0.638	0.636	0.634
			0.648	0.733	0.796	0.840	0.867	0.882
		0.15	0.680	0.700	0.709	0.711	0.710	0.707
			0.666	0.753	0.820	0.867	0.898	0.916
		0.20	0.732	0.756	0.766	0.769	0.768	0.765
			0.674	0.763	0.832	0.882	0.916	0.936
		0.25	0.771	0.797	0.809	0.814	0.813	0.810
			0.676	0.767	0.837	0.889	0.924	0.945
		0.30	0.797	0.826	0.840	0.846	0.846	0.843
			0.676	0.767	0.837	0.890	0.926	0.949
		0.35	0.812	0.844	0.860	0.867	0.868	0.866
			0.673	0.764	0.835	0.888	0.925	0.948

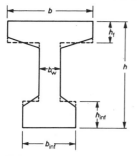

Section modulus ratios	$\dfrac{Z_{inf}}{bh^2/6}$
	$\dfrac{Z_{sup}}{bh^2/6}$

$\dfrac{b_w}{b}$	$\dfrac{b_{inf}}{b}$	$\dfrac{h_{inf}}{h}$	$\dfrac{h_f}{h}=0.10$	$\dfrac{h_f}{h}=0.15$	$\dfrac{h_f}{h}=0.20$	$\dfrac{h_f}{h}=0.25$	$\dfrac{h_f}{h}=0.30$	$\dfrac{h_f}{h}=0.35$
0.35	0.90							
		0.10	0.630	0.648	0.656	0.657	0.656	0.653
			0.655	0.740	0.804	0.849	0.878	0.893
		0.15	0.705	0.726	0.735	0.737	0.736	0.733
			0.673	0.762	0.829	0.878	0.910	0.929
		0.20	0.762	0.786	0.797	0.801	0.800	0.797
			0.681	0.772	0.842	0.893	0.928	0.949
		0.25	0.803	0.831	0.844	0.849	0.848	0.846
			0.683	0.775	0.847	0.899	0.936	0.958
		0.30	0.831	0.862	0.877	0.883	0.884	0.881
			0.682	0.774	0.846	0.900	0.937	0.961
		0.35	0.847	0.881	0.899	0.906	0.907	0.905
			0.680	0.772	0.844	0.898	0.936	0.960
	0.95							
		0.10	0.649	0.667	0.675	0.676	0.675	0.672
			0.661	0.748	0.813	0.858	0.888	0.904
		0.15	0.730	0.751	0.761	0.764	0.762	0.759
			0.680	0.770	0.838	0.888	0.921	0.941
		0.20	0.791	0.817	0.829	0.832	0.831	0.828
			0.688	0.780	0.851	0.903	0.939	0.961
		0.25	0.835	0.864	0.879	0.884	0.884	0.881
			0.690	0.783	0.855	0.909	0.947	0.971
		0.30	0.865	0.897	0.914	0.921	0.921	0.919
			0.688	0.782	0.855	0.910	0.948	0.973
		0.35	0.881	0.917	0.936	0.945	0.946	0.944
			0.686	0.779	0.852	0.907	0.946	0.972
	1.00							
		0.10	0.667	0.686	0.694	0.696	0.694	0.691
			0.667	0.755	0.821	0.867	0.898	0.915
		0.15	0.755	0.777	0.787	0.790	0.788	0.785
			0.686	0.777	0.847	0.898	0.932	0.953
		0.20	0.821	0.847	0.860	0.864	0.863	0.860
			0.694	0.787	0.860	0.913	0.950	0.973
		0.25	0.867	0.898	0.913	0.919	0.919	0.916
			0.696	0.790	0.864	0.919	0.957	0.982
		0.30	0.898	0.932	0.950	0.957	0.958	0.956
			0.694	0.788	0.863	0.919	0.958	0.984
		0.35	0.915	0.953	0.973	0.982	0.984	0.982
			0.691	0.785	0.860	0.916	0.956	0.982

Appendix C

Graphical methods for determining the magnitude and position of the prestressing force

I. Magnel's method [7.1]

This method employs a graph on which the ordinate represents $1/P$, the reciprocal of the prestressing force at transfer, and the abscissa represents e, the eccentricity of the prestressing force.

The four stress conditions (7.1) to (7.4) on page 192 combined with the prestress equations yield four linear relationships between $1/P$ and e (equations 7.1(a) to 7.4(a) on page 203). These may be arranged as follows:

$$\frac{1}{P} > \frac{e + Z_{inf}/A}{M_{min} + f_{cp\,adm}\,Z_{inf}}, \qquad\qquad \text{C.1b}$$

$$\frac{1}{P} < \frac{\eta(e + Z_{inf}/A)}{M_d - f_{t\,adm}\,Z_{inf}}, \qquad\qquad \text{C.2b}$$

$$\frac{1}{P} > \frac{e - Z_{sup}/A}{M_{min} + f_{tp}\,Z_{sup}}, \qquad\qquad \text{C.3b}$$

$$\frac{1}{P} < \frac{\eta(e - Z_{sup}/A)}{M_d - f_{c\,adm}\,Z_{sup}}, \qquad\qquad \text{C.4b}$$

or where prestress at top is compressive:

$$\frac{1}{P} < \frac{Z_{sup}/A - e}{f_{c\,adm} - M_d}.$$

C.5b

These stress conditions are represented by straight lines on the graph $1/P$ by e as in Fig. C.1, and the possible combinations of $1/P$ and e are represented by points lying within the quadrilateral formed by the four lines. Figure C.1 shows the diagram for a beam in which the prestress at the top is tensile, and Fig. C.2 is for a compressive prestress at the top. In the latter case equation C.3b, representing the stress condition relating to the permissible tensile stress at the top at transfer, is no longer relevant.

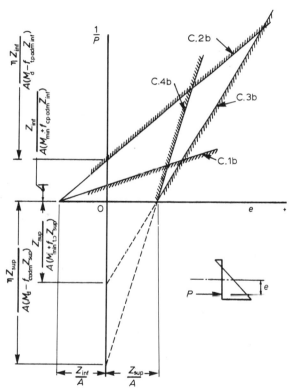

Figure C.1   Magnel's diagram for magnitude and position of prestressing force – tensile prestress at top

II. Author's method [7.2]

The basis of this method is a simple geometrical construction by which the prestress distribution diagram can be drawn for any given values of P and e, the magnitude and eccentricity of the prestressing force, in a section of known

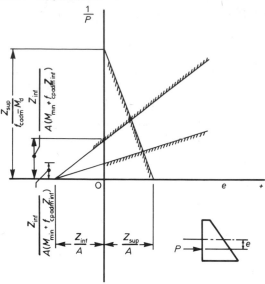

Figure C.2   Magnel's diagram for magnitude and position of prestressing force – compressive prestress at top

dimensional properties. Alternatively the values of P and e may be graphically determined so as to give a required prestress in the concrete.

In Fig. C.3 let A be the area of the section y_{inf} the distance of the centroid from the bottom, i the radius of gyration, and P the prestressing force acting at a distance e, below the centroid. To construct the diagram of the distribution of the prestress draw AB representing the depth of the section; G represents the position of the centroid and S the position of the prestressing force. Draw GC to represent P/A to a suitable scale of stress and GQ to represent i to the linear

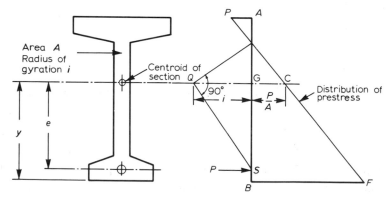

Figure C.3   Graphical construction for magnitude and position of prestressing force

scale. Obtain the point O on AB (produced if necessary) such that angle SQO is 90°. Then O and C are points on the stress distribution line DOCF.

PROOF: Since Q lies on the circumference of a semicircle of diameter OS, $OG \cdot GS = QG^2$. Therefore $OG = QG^2/OG = i^2/e$ and

$$OB = OG + GB = \frac{i^2}{e} + y_{inf} .$$

Also, by similar triangles,

$$BF = GC \cdot \frac{OG}{OG} = \frac{P}{A}\frac{i^2}{e} + y_{inf}\frac{e}{i^2} = \frac{P}{A}\left(1 + \frac{y_{inf}e}{i^2}\right)$$

which is the stress in the concrete at the bottom of the beam. Similarly AD may be shown to be equal to the stress at the top.

When designing the prestress for a section of given dimensions it is convenient to calculate the upper and lower limits of the prestress in the concrete from equations 7.1' to 7.4' on page 197 and to draw the upper and lower boundaries within which the prestress distribution diagram must lie. The above construction then enables the range of possible positions and values of the prestressing force to be examined rapidly, as illustrated in Fig. C.4.

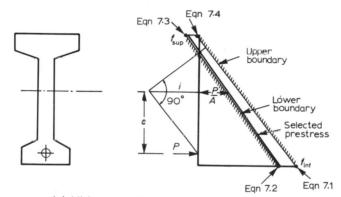

(a) Minimum prestressing force in lowest position

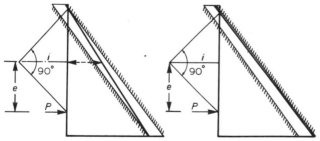

(b) Minimum compressive and tensile stress at transfer

(c) Maximum prestressing force in highest position

Figure C.4 Use of graphical method for selection of prestressing force

Index